OECD Food and Agricultural Reviews

Innovation, Agricultural Productivity and Sustainability in the Netherlands

OECD

BETTER POLICIES FOR BETTER LIVES

Please cite this publication as:
OECD (2015), *Innovation, Agricultural Productivity and Sustainability in the Netherlands*, OECD Food and Agricultural Reviews, OECD Publishing, Paris.
http://dx.doi.org/10.1787/9789264238473-en

ISBN 978-92-64-23845-9 (print)
ISBN 978-92-64-23847-3 (PDF)

Collection: OECD Food and Agricultural Reviews
ISSN 2411-426X (print)
ISSN 2411-4278 (online)

The statistical data for Israel are supplied by and under the responsibility of the relevant Israeli authorities. The use of such data by the OECD is without prejudice to the status of the Golan Heights, East Jerusalem and Israeli settlements in the West Bank under the terms of international law.

Photo credits: Cover © GroentenFruit Bureau

Corrigenda to OECD publications may be found on line at: *www.oecd.org/about/publishing/corrigenda.htm*.

Foreword

Innovation, Agricultural Productivity and Sustainability in the Netherlands is part of the OECD Food and Agricultural Reviews series. It was undertaken at the request of the Dutch Ministry of Economic Affairs. The review examines the conditions in which businesses in the Netherlands undertake innovation in the food and agriculture sector to become more productive and environmentally sustainable. It starts with an overview of the food and agriculture sector and outlines development challenges and opportunities (Chapter 2). A wide range of policies which influence incentives for innovation are then examined: economic stability, governance and trust in institutions (Chapter 3); a favourable and predictable environment for investment (Chapter 4); capacities and public services enabling business development (Chapter 5); agricultural policy (Chapter 6) and the operation of the agricultural innovation system (Chapter 7).

Country policies are analysed following a framework developed by the OECD as part of its work on agricultural innovation and in response to a request from the G20 in 2012 under the Presidency of Mexico. In the first test phase, the framework has been applied to Australia, Brazil and Canada. In this fourth review, specific efforts were made to strengthen the analysis of environmental issues in Dutch agriculture. Additional countries will be studied in subsequent work and the framework is being continuously revised and improved, in particular to reinforce the coverage of sustainability and structural adjustment issues.

This review was prepared by Catherine Moreddu and Julien Hardelin, with contributions from Shingo Kimura. Lihan Wei provided statistical support. Hélène Dernis from the OECD Science, Technology and Innovation Directorate and Douglas Lippoldt, formerly of the OECD, provided data and expertise on intellectual property protection. Martina Abderrahmane provided editorial assistance and Michèle Patterson editorial and publication support.

The review draws heavily on a background report prepared in the context of this activity by Huib Silvis, Jos de Jonge and Jos Verstegen from the LEI Wageningen UR and the Rathenau Institute: "The agricultural knowledge and innovation system of the Netherlands". This material has been complemented by information contained in various OECD publications and databases, and other international databases. An analysis of the determinants of productivity growth in dairy farms was carried out in co-operation with the OECD Farm-Level Analysis network.

This report has benefitted from detailed comments from the Ministry of Economic Affairs, the Ministry of Education, Culture and Science, in particular from the active engagement of Carla Boonstra, Gerty Horeman, Jasper Dalhuisen, and Huib Silvis, and from consultations with a wide diversity of stakeholders in the Netherlands. It has also received valuable comments by Ken Ash, Carmel Cahill and Frank Van Tongeren from the OECD Trade and Agricultural Directorate and from Michael Keenan from the OECD Science, Technology and Innovation Directorate.

This review was declassified by the Working Party on Agricultural Policies and Markets in May 2015.

Acronyms

Acronym	English	Dutch
AIS	Agricultural Innovation System	
AKIS	Agricultural Knowledge and Innovation System	
AOC	Agricultural Education Centres	
AWU	Annual Work Unit	
BERD	Business Expenditure on R&D	
CAP	Common Agricultural Policy	
CBS	Statistics Netherlands	Centraal Bureau voor de Statistiek
CITO	Central Institute for Test Development	Centraal Instituut voor Toetsontwikkeling
CMO	Common Market Organisation	
DLO	Agricultural Research Institute(s)	Dienst Landbouwkundig Onderzoek
DLV	Agricultural Extension Service	Dienst Landbouw Voorlichting
DNB	Dutch Central Bank	De Nederlandsche Bank
DUO	Education Executive Agency	Dienst Uitvoering Onderwijs
EAFRD	European Agricultural Fund for Rural Development	
EAGF	European Agricultural Guarantee Fund	
EC	European Commission	
ECB	European Central Bank	
EQS	Environmental Quality Standards	
ESI	European Structural and Investment (fund)	
EU	European Union	
EU15	15 member states of the European Union, which were members in 2003	
EU12	12 member states of the European Union, which joined after 2003	
EU28	28 member states of the European Union in 2015	
FAO	Food and Agriculture Organization	
FAS	Farm Advisory System	

FDI	Foreign Direct Investment	
FP7	Seventh Framework Programme for Research and Technological Development	
GCI	Global Competitiveness Index	
GDP	Gross Domestic Product	
GERD	Gross domestic expenditure on R&D	
GHG	Greenhouse gas	
GNB	Gross Nitrogen Blance	
HA	Hectare	
HAVO	Senior general secondary education	Hoger Algemeen Voortgezet Onderwijs
HBO	Higher professional education	Hoger Beroeps Onderwijs
	Universities for applied sciences	
HERD	Higher-education Expenditure on R&D	
IBO	Interdepartmental policy evaluation	Interdepartementaal Beleidsonderzoek
ICT	Information and Communications Technology	
ILG	the Rural Area Investment Budget	Investeringsbudget Landelijk Gebied
IP	Intellectual Property	
IPR	Intellectual Property Rights	
JPI	Joint Programming Initiatives	
KIC	Knowledge and Innovation Communities	
LEI	Dutch: Agricultural Economics Research Institute	LEI Wageningen UR
LTO	Federation of agriculture and horticulture	Land- en Tuinbouw Organisatie
MBO	Secondary vocational education	middelbaar beroepsonderwijs
MFF	Multiannual Financial Framework	
MFN	Most-Favoured Nation	
MINAS	Mineral Accounting System	
MIT	SME Innovation Stimulation Top sectors	MKB-innovatiestimulering Topsectoren
MKB	Small and Medium Sized Enterprise (SME)	Midden En Klein Bedrijf
N	Nitrogen	
NRLO	National Council for Agricultural Research	Nationale Raad voor Landbouwkundig Onderzoek

NVWA	The Netherlands Food and Consumer Product Safety Authority	Nederlandse Voedsel- en Warenautoriteit
NWO	Foundation for Scientific Research	Nederlandse Organisatie voor
OECD	Organisation for Economic Co-operation and Development	
OVO	Research-Extension-Education	Onderzoek-Voorlichting-Onderwijs
PCT	Patent Co-operation Treaty	
PMR	Product Market Regulation	
PO	Producers Organisation	
R&D	Research and Development	
RDA	R&D allowance	
RDP	Rural Development Plan/Programme	Plattelandsontwikkelingsprogramma (POP)
RDR	Rural Development Regulation	
RIVM	National Institute for Health and Environment	Rijksinstituut voor Volksgezondheid en Milieu
ROC	Regional training centre	
RVO	Netherlands Enterprise Agency	Rijksdienst voor Ondernemend Nederland
SCAR	Standing Committee on Agricultural Research	
SME	Small and Medium Sized Enterprise	Midden En Klein Bedrijf (MKB)
TIFN	Topinstitute for Food and Nutrition	
TiVA	Trade in Value-Added	
TFP	Total Factor Productivity	
TKI	Top consortium for knowledge and innovation	
TNO	Netherlands Organisation for applied scientific research	Nederlandse Organisatie voor Toegepast Natuurwetenschappelijk Onderzoek
TTI	Technological Top Institute	
TWIN	Total Investment in Research and Innovation	Totale Investeringen in Wetenschap en Innovatie
UAA	Utilised Agricultural Area	
UWV	Employee Insurances Implementing Agency	Uitvoering WerknemersVerzekeringen
VMBO	Pre-vocational secondary education	Voorbereidend Middelbaar Beroepsonderwijs
VSNU	Association of (co-operating) universities	Van Samenwerkende Nederlandse Universiteiten
VWO	Pre-university secondary education	Voorbereidend wetenschappelijk onderwijs

WBSO	Payroll tax allowance	
WEF	World Economic Forum	
WO	University	Wetenschappelijk Onderwijs
WRR	Scientific council for government policy	Wetenschappelijke Raad voor het Regeringsbeleid
WTO	World Trade Organisation	

Table of contents

Tables

Figures

Executive summary

The Dutch food, agriculture and horticulture sector is innovative and export oriented, with high value-added along the food chain and significant world export shares for many products. Continuous adoption of innovation has permitted to reach high levels of productivity, and sustained productivity growth, in particular at the farm-level, in a context of increasing environmental regulatory constraints. The challenge is whether marginal improvements in current technologies and know-how will be enough to pursue current rates of productivity growth, sustainably, and whether the innovation system will be able to generate the new ideas that are likely needed to face future challenges, including those linked to climate change.

The overall policy environment is one of the most favourable to investment worldwide, including for innovation to increase productivity and sustainability. The Netherlands is a highly developed and knowledge-based economy, which benefits from the EU Common market and its 500 million consumers, and is exposed to world markets, with trusted and well-functioning institutions, and generally sound policies. The economy is gradually recovering from the adverse effects of the financial and economic crisis, and budget deficits have been reduced. The policy environment is favourable to innovation because of the ease of doing business, well-functioning and competitive markets, openness to trade and investment, which also facilitates knowledge transfers, high-quality infrastructure, and high-quality education systems responsive to business demand providing a well-educated and skilled labour force.

But there is scope for further improvement in some areas. Difficult access to capital, especially for small and medium-sized enterprises (SMEs) since the crisis may limit innovation investment in agri-food. Government support to private investment targets the adoption of innovation, but simplification of procedures would facilitate access to public support. Tax incentives linked to R&D labour costs and to corporate profits are increasingly important instruments to support innovation activities in private firms, but smaller agri-food companies, with low profit margins and little or no research capacity, cannot take advantage. They can, however, benefit from direct support for the adoption of technological and non-technological innovation. Collaborative efforts are being made to match skills, but a better anticipation of the growing and more diverse demand for skills and increased flexibility in the labour market and immigration policy would help. Finally, the business environment for innovation would benefit, when feasible, from simpler, shorter procedures, which would reduce regulatory transaction costs for business, and from regulatory systems that are more reactive to new types of innovation.

The Dutch government generally opts for less distorting and most efficient agricultural policy options within the Common Agricultural Policy (CAP) context. Commodity-specific payments are kept to a minimum. CAP rural development funds are concentrated on a small number of measures with clear objectives to increase impact. They include support for the adoption of animal welfare and environmentally-friendly practices, including the management of manure from livestock production. Investment assistance is usually targeted to improvements needed to respect regulations and to innovate. Until recently the CAP also supported general rural activities. The recent CAP reform offers new opportunities: the removal of quota will strengthen adjustment to market signals, but environmental and other constraints may limit production growth, while encouraging wider adoption and development of innovative solutions to overcome these constraints; greening measures in the CAP may facilitate improvements in environmental performance, in particular biodiversity. It is crucial for agricultural policy to provide a long term vision for the entire agri-food sector, which recognises the need to improve environmental performance while maintaining productivity growth.

The Dutch agricultural innovation system is a high performer at the national and international levels. It is characterised by strong supply of innovation despite decreasing funding; a demand driven agenda and good collaboration between research, education and industry within sectors, which ensures innovations are relevant and widely adopted at the farm and agri-food firm levels; high participation in international collaborative efforts, and in particular at the EU level. Agri-food innovation benefits from high quality education and research institutions and competitive agri-food industries. Wageningen UR facilitates pluridisciplinary approaches among agri-food experts (e.g. economic, science, agricultural and natural resources), but less collaboration with other specialists.

Developments in innovation policy increase uncertainties. Sources of funding have become more uncertain, with the reduction in public expenditure, in particular for research and education institutions, the increasing share of project-based competitive funding and the higher dependence on EU funds. The abolition of product board levies and the growing role of tax incentives for research and development (R&D) in support of private innovation also reduce resources available for applied agri-food R&D. Given the lag between investment in R&D and results it is difficult to know whether current investments will be enough to maintain long-term performance, including the capacity to collaborate at the international level. The government has traditionally played a strong role in guiding and funding R&D investment. Now, the agenda of the Top Sector policy, which provides a strategic framework for innovation policy, is jointly established by the private sector, the knowledge institutes and the government, with the private sector playing a more important role, as it is a full partner of the public sector in its funding and implementation. There are questions as to how investments in R&D with strong public goods aspects and for long-term challenges such as climate change will be met in a system driven largely by industry. Finally, most public funding is channelled through public-private partnerships and it remains to be seen whether their design is always an efficient way of spending public money. The performance of the current arrangements will need to be carefully monitored as pursuing longer-term challenges usually requires strong government strategic steering and public-private partnerships require strong government leadership to ensure effective use of public money.

Policy recommendations encompass the following four key areas:

- ***Improve further incentives for private investment*** including by minimising the transaction costs of compliance to regulations, for registering new products, and improving the architecture of investment support programmes, in particular by revisiting tax incentives and investment support programmes.
- ***Improve further capacities and services for innovation,*** including by better anticipating future demand for skills, facilitating labour mobility and on-the-job training, strengthening linkages and breaking institutional boundaries between "green" and general education funding to ensure equal access.
- ***Strengthen agricultural policy incentives to innovation for sustainability and longer-term challenges,*** by developing a longer-term vision reconciling productivity growth and sustainability; continuing to provide information on current and future opportunities and challenges, increasing further the targeting of CAP rural development programmes towards support for the adoption of innovative practices; improving the capacity of farmers to participate in the agricultural innovation system (farm advisory, producer groups, agri-environmental incentives); and revisit the existing mix of regulation, financial incentives, and innovative market-based mechanisms to improve the preservation of natural resources and foster eco-innovation, i.e. innovation that is less environmentally harmful than relevant alternatives.

Strengthen the long-term performance of the food and agricultural innovation system, by reinforcing the role of the government in shaping the research agenda to improve the consideration of longer-term and public good issues; by including longer-term impacts in policy evaluation; by introducing mechanisms to better reflect societal demand and foster investment in public goods and long terms challenges such as climate change; by identifying new, more stable sources of funding for longer-term challenges; by improving long-term stability in funding, by dedicating some public investment for knowledge infrastructure and institutions, and long-term challenges; by continuing to monitor and evaluate innovation adoption, by including of environmentally-friendly practices; and by strengthening the links between agriculture-specific innovation systems and related areas (health, environment).

Chapter 1

Overall assessment and recommendations

This chapter presents the framework used in the report to analyse the extent to which Dutch policies are supportive of innovation and structural change, and affect access to and use of natural resources for productivity and sustainability, and an overview of the findings of the review of a wide range of policies in the Netherlands. In each policy area, it develops specific policy recommendations.

Improvements in agriculture productivity growth are required to meet the growing demand for food, feed, fuel and fibre, and must be achieved sustainably through the more efficient use of natural and human resources. A common finding is that a wide range of economy-wide policies affect the performance of the food and agriculture sectors, and thus need to be considered alongside agriculture-specific policies. Recognising that innovation[1] is essential to improving productivity growth sustainably along the whole agri-food chain, OECD work dedicated specific attention to the performance of agricultural innovation systems.

The framework used in this report to review Dutch policies considers policy incentives and disincentives to innovation, structural change and access to natural resources, which are key drivers of productivity growth and sustainable use of resources (Figure 1.1). The focus was originally on agricultural innovation and productivity, with on-going efforts to strengthen sustainability aspects reflected in this report.

Figure 1.1. Policy drivers of innovation, productivity and sustainability in the food and agriculture sector

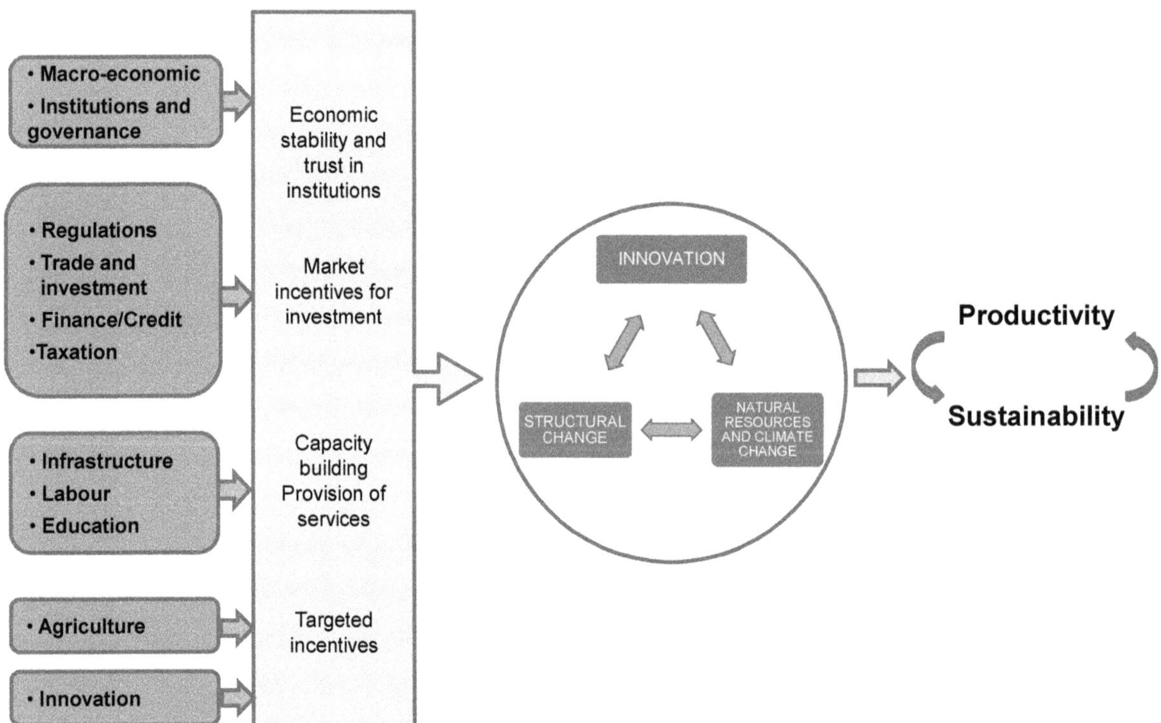

Source: OECD (2014a), "Analysing Policies to improve agricultural productivity growth, sustainably: Revised framework". http://www.oecd.org/tad/agricultural-policies/innovation-food-agriculture.htm.

This review begins with an overview of the characteristics and performance of the food and agriculture sector and the challenges it will face in the future (Chapter 2). A wide range of policies is then considered according to the main channels or incentive areas through which they affect drivers of productivity growth and sustainable use of resources.

- Economic stability and trust in institutions (justice, security, property rights), both of which are essential to attract long-term investment in the economy (Chapter 3).

- Private investment, which in turn requires a transparent and predictable environment that balances the interests of investors and society (Chapter 4).
- Capacity building, including provision of essential public services (Chapter 5).
- Agricultural policy, domestic and trade related (Chapter 6).
- The agricultural innovation system (Chapter 7).

A policy area can affect innovation through more than one channel. Policies can affect innovation positively or negatively depending on the type and intensity of measures. This review draws on background information provided by Dutch consultants and on recent OECD economic, territorial, environmental and innovation reviews (OECD, 2008, 2014b,c,d, 2015).

This report aims to review the extent to which the policy environment contributes to improving productivity growth and sustainable use of resources in the food and agriculture sector by fostering structural change, sustainable use of natural resources and the creation and adoption of innovation. Throughout the report, the likely impacts of each policy area on innovation are first discussed in general terms. Specific country measures are then analysed in this regard. Overall assessment and recommendations are drawn from this review on a large range of policy areas.

Overview of the Dutch food and agriculture sector

The Dutch food, agricultural and horticultural sector has several strengths including natural and geographical conditions favouring diverse agricultural activities, a resilient primary production structure of family enterprises and well-educated labour force, integration in net chains (networks and supply chains) of agro-food products and a strong international orientation. The high share of agro-food exports in total exports and in GDP is partly based on its high levels of land and labour productivity in primary agriculture driven by widespread and continuous adoption of innovation that has increased input use efficiency in recent years, and on the composition of its high value production package, resulting from a continuous process of rationalisation, consolidation, mechanisation and specialisation. This process has been supported by government policies that focused on agricultural development in a general environment conducive to innovation.

At the same time, the sector faces relatively high land and labour costs as is expected in a wealthy economy with a high population density. Moreover, conditions for the sector become increasingly uncertain with economic risks linked to price variability and concentration of operations rising; environmental and animal welfare constraints increasing as government regulations and policy respond to society's demands; and the nature and magnitude of the impact of climate change at local level remaining largely unknown.

For the Dutch agri-food complex to respond to growing demand for food at the world level and for diversified products, with quality attributes in high-income markets, the agricultural innovation system will have to supply solutions that help save resources and energy, while improving the productivity and quality of agri-food products. Agricultural policy could then foster adaptation of the sector to its changing environment, by removing remaining market distortions and increasing the scope for targeting payments to specific objectives such as investment in innovative technologies, adoption of sustainable practices or risk management tools.

Improve further the supportive framework conditions for innovation and entrepreneurship

The economic crisis has affected both public and private capacity to invest in innovation. The Netherlands is a high-income, trade-oriented and knowledge-based economy, with sound policy and efficient institutions. The world financial and economic crisis has, however, hit severely this open economy, which is gradually recovering from a prolonged recession. Fiscal consolidation and structural reforms have achieved a reduction of budget deficits to less than 3% of GDP, as required by the EU

Stability and Growth Pact. This situation has consequences on the resources for innovation both via the reduction of government expenditure and the difficult access to capital for businesses, in particular small and medium-sized enterprises (SMEs) with lower profit margins.

Reforms have significantly reduced regulatory barriers to entrepreneurship, but there is still scope for reducing complexity and transaction cost related to compliance with regulations. Overall, regulatory barriers to entrepreneurship are the least restrictive amongst OECD countries, and businesses operate in a competitive environment conducive to investment. Low administrative burdens on start-ups in particular favour the development of innovative activities. There is, however, scope for further improvement. Regulations remain relatively complex and costly, although much less than the OECD average, and the government has set targets to reduce administrative burdens and compliance costs for enterprises, and improve transparency and provision of public services. Specific areas where further efforts would be required to place the Netherlands among the top 5 OECD performers are the simplification of the licence and permits system, the reduction of administrative burdens for corporations and barriers in services and network sectors, and the lowering of legal barriers to entry.

Environmental regulation has become more stringent over time, but additional incentives would improve compliance. Environmental regulation stringency in the Netherlands is among the strictest of OECD countries. Combined with other incentives, this has led to significant improvements in the environmental performance of agriculture, with decreasing trends in terms of nutrient surplus and pesticide use by the sector. To some extent, environmental constraints have driven innovations needed to comply. However, despite such encouraging trends, compliance with EU environmental directives is still imperfect and progress in agri-environmental performance tends to slow down, or even worsen in the case of biodiversity. The ability of agriculture to reach environmental targets could become among the most important challenge for growth of output and productivity in the sector in the future, thus making the role of eco-innovation central in that regard.

Competition for land use between agriculture and other activities has increased with the decentralisation of land use policies, the focus of regional policy on promoting economic development, and move away from restricting land use.

The impact of EU "regulations on products and processes" on innovation is being reviewed, with independent evaluation being commissioned on several legislative areas concerning agriculture and food. The effort to make regulations smarter aims to simplify existing legislation, minimise administrative compliance costs and reduce the cost of registering new products and processes. In this context, particular attention should be paid to regulations on processes, which can be an obstacle to innovation.

Efforts to reduce unnecessary administrative burden should continue. For the food processing sector, EU regulation can be complex and open to interpretation. This can be an issue, in particular for the SMEs. A variety of innovative approaches can help reduce regulatory costs, without compromising outcomes. Moreover, it is important that regulatory procedures keep up pace with innovation (food safety, novel food).

The trade and investment environment facilitates knowledge flows embedded in agri-food trade and Foreign Direct Investment (FDI). Restrictions to trade and investment are the lowest among OECD countries and emerging economies, although some agricultural products are protected by EU common tariffs and tariff quotas. The country is open to FDI, in particular in agri-food as illustrated by investments by Fonterra, but difficult access to credit for foreign companies limits FDI. The Netherlands is a significant recipient of agri-food FDI, but it is also a net investor abroad. Agri-food FDI flows include knowledge transfer, which is seen as an embedded traded service, but they are also complementary to trade as a means to ensure traceability and quality of foreign products exported to the Netherlands. Market opportunities for transferring Dutch know-how are growing in countries with similar small farm structure, such as China.

Access to finance for innovative firms has decreased since the financial crisis has weakened Dutch banks. The Dutch financial market is well-developed and the banking sector is large compared to the economic size of the country and has suffered major losses on international capital markets at the beginning of the financial crisis. Despite progress to strengthen bank capital, exposure to non-performing loans not covered by loan loss provisions is high. The dependence on international capital markets remains large and the volatility of risk premia has increased. In addition to the current weakness of the banking sector, venture capital, which is particularly important for innovative firms, is lacking. The cooperative Rabobank is the main supplier of credit to farmers. Since the beginning of the crisis, Rabobank loans to the farming sector have remained stable, with diverging trends by commodity sub-sector, mainly related to differences in economic performance and opportunities: dairy farmers borrowed more to invest in the preparation of the abolition of the dairy quota, while the horticulture sector received less credit due to lower margins. There are multiple programmes to support investment, some being targeted to different stages of innovation. This tool box is favourable to expansion of existing firms, but is found complicated and difficult to navigate through for new firms, which do not have specialists to help them, all the more because programmes change frequently. To address this issue, the Netherlands Enterprise Agency (RVO), and a single desk platform launched in 2014 (*ondernemersplein*) provide information and advice to entrepreneurs for free. Stakeholders find it relatively easy to get funding for prototypes but difficult for bringing a product to the market. These investment support programmes benefit upstream and downstream industry, and, within EU state aid conditions, primary agriculture.

Credit support under Pillar 2 of the CAP is generally targeted to investments to improve competitiveness and sustainability, in particular compliance with environmental, food safety and animal welfare regulations.

Tax rates on personal and corporate income are close to the OECD average or median. The difference between the preferential rate for smaller profits and the standard rate is too small to act as a barrier to business growth. Few tax rebates are specific to agriculture, although agriculture benefits from tax rebates on green subsidies or investments, and tax exemptions on land transfers or capital gains on land. Glasshouse horticulture benefits from a lower energy tax. The Diesel tax rebate for agriculture uses was abolished on 1 January 2013.

Tax incentives for innovation have increased in recent years and ***account for over three-quarters of government support to business innovation.*** They benefit mainly sectors with high profit margins and companies with employees involved in Research and Development (R&D). Tax incentives are used in horticulture, but not so much by small food processing companies, which do not have the capacity to carry out R&D activities.

Recommendations to improve incentives for private investment

- Efforts to minimise administrative costs of compliance and reduce the costs of registering products, and reduce length and simplify procedures, need to continue. Regulators need to keep up pace with innovation (food safety, novel food) and when possible, avoid regulation on processes that hinder future innovation. Focus on the reduction of administrative burdens for corporations and barriers in services and network sectors, and the lowering of legal barriers to entry to strengthen competition.
- Focus public support to investment in areas where financial markets fail to provide funds. Continue efforts to help the banking sector regain its former strength. Simplify the architecture of credit support programmes to improve access and targeting.
- Rebalance the policy mix by complementing the current focus on R&D tax credits with competitive, well-designed direct support instruments, e.g. for joint R&D projects with knowledge institutes, and instruments used in the top sectors approach, such as the SME Innovation Stimulation Top sectors (MIT) (Annex 1.A1).
- Foster stability and minimise the burden imposed on businesses by frequent changes in the policy mix. Predictability could be improved by linking major policy changes to system evaluation cycles agreed upon in advance (e.g. over five-year periods) (Annex 1.A1).

Improve the capacities and services for innovation

Economic activities and rural populations benefit from an ***excellent infrastructure network and good access to public services.*** The infrastructure network is one of the world's best, in particular port infrastructures. Public services and economic and social conditions are similarly attractive in all regions.

More flexible employment and migration policy would help match skills needs. The Dutch legislation protects employees, although some provisions have been recently relaxed. Given the importance of horticulture, there is a strong demand for seasonal work, which has been increasingly met by temporary migrants. Seasonal migrant workers have become essential for the survival of the sector.

Education and training can help meet the growing demand for high-skilled labour. Dutch agriculture is technologically advanced, in particular the horticulture sector where innovation is an on-going process leading to rapid circulation of knowledge and technologies. As a result there is in some sectors a strong demand for skilled workers and for upgrading skills. This demand is being addressed in collaboration with the education system, but delays in the response may lead to temporary shortages of skills. The extent to which subsidies to train employees and innovative teaching methods to reach a wider public could help improve the skills base for the sector should be investigated.

Increasing linkages between agricultural and general education would benefit innovation in food, agricultural and horticulture, by increasing the knowledge-base of 'green' and 'non-green' students to include common areas of interest, facilitating knowledge transfer across areas through students' movements, and facilitating the adaptation of general knowledge to sector-specific issues. The Dutch education system is highly-ranked at the world level and it is responsive to business needs. As a result, the population and labour force are well-educated, with above-average proficiency in literacy, numeracy and problem solving in technology-rich environments, and as well as high frequency in using information and communication technology (ICT) at home. This contributes to facilitating innovation in firms and acceptance of innovation in the society. The education system offers strong and attractive degrees in agriculture, food and nature management, concentrated in specialised institutions under the Ministry of Economic Affairs, in charge of agriculture. "Green" education enjoys good branding and offers scope for development, in particular in partnership with the private sector. As a result, the number of students enrolled in agricultural technical and higher education increases, but this growth may be limited by infrastructure capacity and limitations to funding by Ministry of Economic Affairs, as funds for general education cannot be transferred into this growing area. Similarly, as a result of the concentration of agriculture-related activities into specialised institutions, the links with other general education institutions are weaker than they could be otherwise. This concentration facilitates multidisciplinary approaches with agricultural specialists, but it limits exchanges with specialists from other domains, which are increasingly important for agriculture to be able to deal with current and future challenges such as climate change (e.g. human and animal health, energy, water management, ICT, nanotechnology).

Maintaining collaboration with the industry is essential to meet future skills needs of the sector and facilitate its development. Co-ordination between research, education and industry has been long standing and is being strengthened in the "Green Table". Moreover, the good knowledge-base in agriculture-related topics attracts multinationals, which invest in facilities and projects with education and knowledge institutions. The Top Sector policy, which provides a strategic framework for innovation policy, raises concerns in education and knowledge institutions about future funding from the private industry. A Human capital agenda is included in the Top Sector policy but private funding for education focuses on Wageningen University and is not so easy to access for vocational education. In addition, areas not included in the top sectors also have difficulties to attract private funding.

Recommendations to improve capacities and services for innovation

- Increase the flexibility of employment and migration policy to facilitate labour force moving into areas with strong demand, such as agri-food and nature management.
- Ensure public funding for education and knowledge institutions to enable them to continue to offer relevant education and training, and participate actively in the agricultural innovation system. In particular, whatever the ministry in charge, public resources for education should be equally distributed on the basis of the number of students in order to enable students to move to areas with attractive employment prospects such as agri-food education.
- Facilitate discussion between education and knowledge institutions and the industry to identify current and future skills for the development of the sector and the improvement of productivity and sustainability performance. Find innovative ways to improve systems' reactivity to new demand by facilitating further life-long learning and upgrading of skills in the labour force.
- Continue to develop business management programmes, including for future researchers and farmers, to facilitate the valorisation and adoption of knowledge. Learning how to deal with uncertainty and cope with problems will become an ever more important asset.

Strengthen further agricultural policy incentives for the adoption of innovation

Dutch implementation of the CAP generally aims to facilitate productive investment. Within the framework of the EU Common agricultural policy, the Netherlands generally opted for measures that least distort market signals for the agricultural sector and the rural economy. For example, the government chose to remove most payments linked to commodity production when they had the flexibility to do so. For 2014-20, commodity-specific payments, possibly in the form of a grazing premium, a scheme for the sustainability and transition of the veal sector, and a potato starch payment, would amount to less than 0.5% of Pillar 1 direct payments. A characteristic of Dutch rural development programmes under the CAP (Pillar 2) is the concentration of funds on a limited number of measures to reinforce impact. In the programme for 2007-13, a relatively large share of funds was allocated to non-sectoral measures that favour the rural economy. For the programming period 2014-20, the emphasis is placed on innovation, sustainability and nature preservation. Moreover, the government's plan to transfer 10% of Pillar 1 direct payments to Pillar 2 from 2015 will provide opportunities to strengthen funding in these areas.

It is crucial for agricultural policy to provide a long term vision for the sector, which recognises the need to improve environmental performance while maintaining productivity growth. The abolition of EU dairy quotas will offer opportunities to Dutch farmers to increase the scale of operations. This will facilitate further investment in large-scale, innovative technologies needed to remain competitive and improve environmental performance. But the extent to which Dutch farmers can take advantage of the removal of quotas to increase production may be limited by environmental and other constraints. The need to overcome these constraints is likely to encourage wider adoption and development of innovative solutions. Most of EU funding remains in Pillar 1 direct payments, which raise farmers' capacity for investment but slow competition and structural adjustment by maintaining inefficient farm operations in the sector, and increasing assets value and entry costs. Implementation based on historical entitlements, in particular, created inequalities and favoured intensive types of agriculture. Convergence of payments rate per hectare in the Netherlands will reduce support to intensive livestock production units. If farmers adjust to lower support to intensive production systems by adopting less intensive practices or productions, for which they could receive conditional payments and investment assistance, this could lead to lower pressure on the environment. However, the reform of direct payments will reinforce the link between payments and land. Whether it will affect land value depends on changes of the extent of capitalisation of support before and after the reform. Linking 30% of Pillar 1 payments to specific land management practices is expected to increase areas covered by good environmental practices above those already covered by environmental schemes. But the impact on environmental performance remains to be seen.

The information base and analytical tools to monitor progress, evaluate agricultural and innovation policies and guide farmers' decisions should be maintained and even developed. The government has an important role to play in the collection of information, which allows for the formulation of evidence-based policy, improved through monitoring and evaluation. It is particularly important to identify the determinants of the adoption of specific types of innovation and to strengthen the capacity of farmers, or farmers' organisations, to formulate their needs, and participate in knowledge networks.

Recommendations to make agricultural policy more conducive to innovation

- Develop a long-term vision reconciling productivity growth and sustainability and reduce policy uncertainty.
- Continue to limit the provision of coupled payments to very targeted and temporary measures to improve traceability and sustainability, through innovative investments and tools.
- Identify market failures in credit and land markets to design better targeted policies to facilitate investment and farm transfer.
- Strengthen the ability of agricultural policy to improve the environmental performance of agriculture, by focusing agri-environmental measures to objectives and outcomes rather than on process and meet EU regulation constraints; revisit the balance between regulation and economic incentives in view of fostering environmentally-friendly innovation, building on the analysis of the pros and cons of the Dutch experience in this area, such as the Mineral Accounting System (MINAS).
- Make use of the opportunity given by the CAP to recognise Producer and Branch Organisations and support the participation of farmers or farmers' organisations in knowledge networks.
- Maintain the good information base and analytical capacity to monitor progress, evaluate policies and guide farmers' decisions, with specific attention to innovation adoption and environmental practices.

Maintain the strength of the agricultural innovation system

The Dutch agricultural innovation system is a very good performer at the national and international levels. Over the years, it has supported strong growth in domestic food and agriculture productivity, with total factor productivity of primary agriculture being one of the highest at the EU level. It has also developed world leadership in productivity-enhancing technologies, in particular for the greenhouse sector. In the last two decades, it has also developed technologies and practices to improve the efficiency of input use, including natural resources. This is reflected in changes in the sources of Total Factor Productivity (TFP) growth, which originally came from output growing faster than input use, to become mainly driven by the reduction of input use while maintaining or increasing production. The good performance of the agricultural innovation system came from strong investment over the long term, and tripartite collaboration between education, research and industry. Both have led to high R&D supply, accumulation of knowledge stock, and good knowledge infrastructure, allowing for both international collaboration, and the development of solutions adapted to the sector's demand.

Investments from private and public sources in agri-food innovation have been in line with the significance of the sector for the Dutch economy. They are likely to have contributed to productivity and income growth in the sector, although evidence for a causal relation is difficult to establish. It is clear, however, that the agri-food innovation institutes (especially Wageningen UR) are highly ranked institutes, and that the Netherlands has a leading position at the world level in agricultural sciences.

As strong performers in agri-food R&D, Dutch research and education institutions are active and successful partners in collaborative efforts on food and agricultural research, in particular at the EU level, but also in global networks and initiatives. As a result, the share of EU funding in the budget of R&D institutions has increased in the last decade. Excellence in agri-food research and education attracts foreign investment from foreign multinationals interested in research collaboration. Wageningen University attracts foreign staff and students. This has positive short and longer term benefits for the international orientation of the system.

As reflected by the high productivity performance of the sector and shown in innovation surveys, adoption of innovation in farms and firms is widespread. Well-educated farmers have access to a diversity of training and advisory services on a wide range of technical, organisational, management and marketing aspects, which have facilitated the adoption of innovation. Government involvement is now limited to the granting of small subsidies for farmers to access services under Pillar 2 of the CAP. Marketing of new ideas is facilitated by government programmes to support investment at different stages of innovation. Intellectual property protection, which is essential to attracting private investment in innovation, is generally high by international standards, although it is only used when benefits outweigh costs.

Institutional developments have made the system more collaborative and demand-driven and have strengthened the role of the private sector in guiding investment. R&D co-ordination and funding mechanisms have facilitated industry-driven projects, public-private partnerships and networking, at the sub-sector and regional levels. Major changes in agricultural research, extension and education include the privatisation of the national extension service, which has been replaced by a diversity of private providers, and the merging of applied research institutes and the agricultural university into Wageningen UR (University and Research Centre) in the late 1990s. This was presented as a "third generation university" with innovation in its mission. Wageningen UR now dominates the system, ensuring good integration between fundamental and applied research and between research and education. This facilitates multidisciplinary approaches among agri-food experts, but does not facilitate co-operation with innovation actors in other sectors, although experts from Wageningen UR participate in collaborative programmes with experts from different institutes or universities. The abolition of commodity boards has, however, reduced the ability of the primary sector and smaller food companies to participate in the system and formulate their demand.

Concerning professional education, the agricultural universities of applied sciences and the agricultural colleges (technical and vocational training) are developing from schools into regional and international knowledge centres. These aim to contribute to lifelong learning, to innovation and to the license to produce. Learning innovation networks with different parties are used as a policy instrument to address systemic coordination issues. Schools are expected to help firms with qualified employees and practical solutions.

Sources of R&D funding have become more uncertain. A high and increasing share of funds is project-based and delivered through competitive mechanisms to increase efficiency and relevance. But this leads to increasing transaction costs and uncertainties: significant time is needed to develop convincing projects, whether they obtain funding or not. Moreover, depending on the selection criteria, this may not facilitate long term, risky undertakings. At the same time, public funds for applied research have decreased and levies raised from commodity sectors have been abolished in 2011, without being replaced. A recent issue discussed below is that public money for applied research requires private co-financing, and is thus limited to areas where the industry has the willingness and capacity to invest. Provincial funds, which draw on receipts from the privatisation of energy companies, remain quite large but will run out. Besides, the selling of Intellectual Property Rights (IPRs) does not generate significant or any money, given the costly and lengthy registering process. The decline of government funding has also increased dependence on foreign funding, in particular from the European Union, which may become more difficult to obtain because of the pre-harvest focus of the current EU programme, and because of the need to find national co-financing. Beside the rising importance of competitive funding, another development is the shift of responsibility for the knowledge base and knowledge infrastructure to shared public-private responsibility.

Tax incentives provide the majority of support to innovation in the private sector. There have also been changes in instruments used to support innovation in the private sector, with the role of tax incentives increasing while direct support declined. This helps private research institutions directly and public research institutions indirectly because it encourages the industry's participation in Public-Private Partnerships (PPPs). However, tax incentives favour sectors with high profit margins or companies with staff engaged in R&D activities. The agri-food sector includes a high number of smaller firms, with little capacity to hire staff engaged in R&D activities.

The current policy approach accentuates further previous trends in R&D funding, raising questions about the sustainability of the system. More precisely, the system's performance, which builds on accumulated knowledge and capacity, may not be able to maintain its top position with an approach mainly driven by industry interests, which tend to be shorter term.

The general rationale for public sector involvement in innovation is market failure in the provision of innovation with long-term horizon and high risks with uncertain returns. Innovations improving long-term environmental performance, for example, have public good aspects. In the agri-food sector, the fragmentation of demand from a large number of small farms, and regional and sector specificities, also explain the large share of public funding in agricultural R&D. Public leadership in governance also helps ensure innovation responds to societal demand. An important role for the government is also to maintain a knowledge infrastructure that benefits all actors.

The top sector policy introduced in 2011 concentrates public funds for innovation on nine export-leader sectors which make up 55% of exports and over 80% of R&D expenditures (but only 30% of GDP). Two top sectors are dedicated respectively to the export orientated agri-food sector and horticulture and propagation materials sector.

The top sector policy subjects the granting of public funding to participation in public-private partnerships (PPPs) within top sectors and gives industry a leading role in setting innovation agendas. This aims to maintain the competitive edge of these sectors, through innovation. While companies participating in top sectors already invested in innovation, public co-funding focussing on pre-competitive research is expected to reinforce their contribution in this area. Early findings suggest companies, including multinational ones, increased investment in pre-competitive research, but that total private expenditures did not increase overall. The policy was also expected to reinforce networking to the benefit of all partners. Networking was already well-developed and the top sectors built largely on existing networks. In the food industry, however, they improved co-operation between the processing and retail levels, as co-operation already existed among other components of the chain. The PPP approach is also to facilitate the marketing and adoption of innovation, and reduce the technological gap between small and large companies through knowledge transfer, as quality systems become more complex. Knowledge gained through PPPs enters more quickly into the public domain. Moreover, IPR does not seem to be an issue for companies as PPPs are for pre-competitive research.

The top sector approach carries some pitfalls for the long-term: business-driven innovation tends to focus on low risk and short term R&D activities, and generally invests less in fundamental and public good related research, even though some large companies may have a longer term approach. Another issue for the long-term competitiveness and sustainability of the economy is the exclusion of non-top sectors, and the reduction of funding for activities such as outreach and education. To address long-term challenges related to food security, environmental problems and natural resource limitations, including the impact of climate change, innovations with strong public good aspects are needed. Those are unlikely to be forthcoming exclusively from private, industry-driven, initiatives. Moreover, the top sector policy introduces an additional weakness in the system, as projects are conditional on private sector participation. OECD work on PPPs suggests that they should not be considered themselves as an objective, but only as a mechanism to share costs and bring mutual benefits in certain circumstances. While they should be facilitated in these cases, they should not be forced in all circumstances. The limited capacity of agri-food SMEs to participate could result in significant public funding flowing to companies that already have large capacities to invest, including foreign multinationals. The extent to which top sectors benefit the whole national economy, including by enlarging the knowledge base and transferring it to other sectors and non-participant companies, should be carefully monitored.

There are also concerns that the government is paying a larger share of the investment than is apparent: while public-private co-financing is 50-50 in the top sectors "agri-food" and "horticulture and propagation materials", the private sector can benefit from investment support and tax incentives for R&D, which means that in the end, private contribution can be brought down to 30%.

The two top sectors "agri-food" and "horticulture and propagation materials" receive a healthy share of all top sector public funds. Wageningen University is successful at obtaining competitive public money for R&D, and its budget increases. However, the top sector contribution to the financing of public research institutions is limited by the effectively lower share of private co-funding and the fact that the private contribution is often in kind or financed through tax incentives. As a result, the top sector policy cannot fully compensate for the decline in public funds in the long term.

A widely shared view amongst agri-food stakeholders is that a major benefit of the top sector stems from the collaboration along the food chain. The participation of food processing SMEs is, however, not clear. Participating companies are mostly large because companies are expected to invest substantially and research is pre-competitive. Participation of industry in some sectors, particularly horticulture, has become more difficult since the abolition of commodity boards, partly because of the heterogeneity of the food sector in terms of size, but also the horticulture sector, by region and whether greenhouse or not. In a sector characterised by many micro-(family)-enterprises with few or none employees, no R&D department and typically low profit margins (and consequently little room for tax deduction), the adoption of innovation in the sector would gain from stronger government involvement in the definition of clear policy goals and instruments, that would facilitate the development of innovation agendas to set R&D priorities and foster the collaboration among firms, and between firms and knowledge institutions.

A recent OECD Review of Innovation Policy in the Netherlands (OECD, 2014d) recommends engaging more small innovative businesses in the "Top Sector" strategy. This strategy focuses public resources on strengthening sectors where Dutch firms already excel. The review also suggests increasing support for business innovation through joint R&D projects with research institutions rather than just through R&D tax credits (see Annex 1.A1). The OECD Territorial Review of the Netherlands recommends aligning the Top Sector innovation strategy with the EU regional cluster policy in order to provide more coherent local incentives.[2] These recommendations also apply for the agricultural innovation system (Annex 1.A1).

More time is needed to evaluate the more fundamental, longer term impact of the top sector on agricultural innovation capacity, but looking from the past to the future there are some worrying signals. First of all there is a tendency of declining public investments generally and for agricultural technical institutes specifically as they are less likely to attract co-funding from the industry. Secondly, fiscal measures have been growing significantly over the past few years and will probably be of vital importance in the coming years. Both the top sector policy and fiscal instruments do not reach agri-food actors as smoothly as the former sector specific investment strategies did. On the other hand, a positive signal is that so far the declining public investments in the recent past (direct and indirect) have been compensated for by rising private investments. Whether or not this trend will continue is hard to predict.

The ambitions of the private sector and the government are changing rapidly, making it difficult for long-term R&D to adjust, and for knowledge institutions to pursue long-term objectives. Public applied research institutions were involved in pre-competitive research for society's benefits. With changes in funding mechanisms, those institutions, in particular the agricultural research institutes, lost some of their ability to establish long-term programming as they need to respond to government and private priorities to obtain funding.

For a number of fields public and private investment will continue, but in others with public good aspects, continuity will be difficult: e.g. food safety (free rider's behaviour), socio-economic and policy-relevant research, research reflecting societal demand and responding to longer term challenges such as climate change and long-term environmental performance of agriculture. It would be important to identify these areas and find alternative solutions to support societal driven innovation. For example, civil society organisations could play a larger role in agricultural innovation systems, or project proposals selection criteria could include a sustainability criteria.

There are also concerns about the capacity of the system to maintain its contribution to international or regional collaborative efforts. On the one hand, the stability of the Dutch system becomes increasingly dependent on EU funding, with the reduction of government expenditure on R&D. On the other hand, competition for obtaining foreign research funding becomes more intensive and transactions costs increase. Moreover, it is increasingly difficult for Dutch researchers to obtain the national co-financing needed for participation in EU projects, as public expenditure on R&D is decreasing in general and in particular in areas not covered by the top sectors, or those of less interest for businesses because of strong public good components (such as policy-relevant activities). Finally, international collaboration requires the maintenance of world-level capacities, which requires basic government support over time.

Further changes in governance and policy design are foreseen. The trend to dilute the special treatment of the agricultural sector in policies, including research and education, is likely to continue. For the Dutch agricultural innovation system, this may lead to further integration with the Federation of applied research institutes[3] in one programme. In the past the agricultural ministry played a leading role in formulating, with the input of stakeholders, the agricultural research programmes rather autonomously, since it had ample resources available. A network approach is now needed to formulate a strategy. This could stimulate synergies and cross-sectoral co-operation.

In the Netherlands, research and innovation programmes are also used as a policy instrument to reach specific public goals (e.g. regarding the environment, animal welfare) and to combine them with other types of regulation. In these areas, there is strong interaction with the primary sector, through farmers' organisations but the interaction with innovation in the food industry may be weak, and it is not very clearly taken into account in the policy design. A point of concern may be that innovation along the food chain is governed by different policies and by multiple policy levels (EU, country). Although incentives might be used to stimulate collaboration, this can easily be disturbed by policy changes in one of the domains, thus the need for better policy co-ordination.

Better information on challenges and opportunities for the sector is essential to guide private investment and policy decisions. The government has an important role to play in providing information systems needed to share information, reduce information gaps to better guide private investment decisions, monitor economic and environmental performance of the sector, identify market and policy failures, and improve policy design, implementation, monitoring and evaluation. Better information and analytic tools are also needed to monitor and evaluate the performance of the whole agricultural innovation system. Individual policies and institutes are regularly evaluated, but so far, there is no systematic mechanism in place to evaluate the agricultural innovation system and the information to do so is fragmented.

Recommendations to strengthen direct incentives to innovation

- Strengthen the role of the government in defining long-term objectives for R&D and innovation, taking into account long-term challenges and societal demand.
- Improve policy co-ordination amongst agricultural, industrial, innovation, education, and regional policies, and policy stability.
- Facilitate the organisation of producers and the industry to enable them to contribute more effectively and efficiently to the agricultural innovation system, including through participation in networks or formulation of demand.
- Strengthen the stability of R&D funding, by dedicating some public investment for the maintenance of knowledge infrastructure and for issues with a longer term horizon.
- Facilitate access to other sources of funding: How could revenues from IPRs be increased? Explore ways to increase IPRs revenues or generate additional funding from royalties or levies.
- Ensure the contribution that business makes to public-private partnerships is commensurate with the benefits they get.

- Identify and fund areas not covered by public-private partnerships, with specific attention to food safety, sanitary and phytosanitary issues, economic analysis, societal issues of no direct interest to the private sector, longer term and more risky issues.
- Consider extending coverage, or at least transferring valuable experience, to other sectors, in particular those with greater scope to improve the intensity, scope and ambition of their innovation activities (Annex 1.A1).
- Explore ways to generate new (breaking through) ideas to overcome current constraints, for example through demand-driven mechanisms, including to develop technologies and systems allowing for a better management of natural resources and improved resilience to risks.
- Ensure public co-financing is available for participation in EU programmes and international collaborative efforts.
- Continue developing information systems, including market intelligence (big data) and research results, as innovation and policy evaluation become more complex and require a wealth of information. In particular, continue to monitor innovation adoption and environmental performance in surveys, in addition to economic performance, to better understand determinants and policy impact. Continue to use and share innovative methods to reduce collection costs and improve farm and firm participation.
- Develop indicators and tools to evaluate the performance of the agricultural innovation systems in general, and innovation policy regularly, taking longer term effects into account, possibly in collaboration with other countries and organisations.

Notes

1. The Oslo Manual defines innovation as the implementation of a new or significantly improved product (good or service), or process, a new marketing method, or a new organisational method in business practices, workplace organisation or external relations (OECD and Eurostat, 2005).
2. Press communiqué http://www.oecd.org/newsroom/netherlands-make-economic-innovation-and-territorial-reforms-work-together-to-boost-growth-and-competitiveness.htm.
3. Through the Federation TO2 six Dutch research organisations for applied research joined forces to deliver added value in the field of applied knowledge. It is the Netherlands Organisation for Applied Scientific Research (TNO), the four Large Technological Institutes (GTIs) – Energy research Centre of the Netherlands (ECN), Maritime Research Institute Netherlands (MARIN), Deltares, and National Aerospace Laboratory (NLR) – and the research institutes of Wageningen UR.

References

OECD and Eurostat (2005), *Oslo Manual: Guidelines for Collecting and Interpreting Innovation Data, 3rd Edition*, The Measurement of Scientific and Technological Activities, OECD Publishing, Paris.
DOI: http://dx.doi.org/10.1787/9789264013100-en.

OECD (2015), *OECD Environmental Performance Reviews: Netherlands 2015*, OECD Publishing, Paris. Forthcoming.

OECD (2014a), "Analysing Policies to improve agricultural productivity growth, sustainably: Revised framework". Available at: www.oecd.org/agriculture/policies/innovation.

OECD (2014b), *OECD Economic Surveys: Netherlands 2014*, OECD Publishing, Paris.
DOI: http://dx.doi.org/10.1787/eco_surveys-nld-2014-en.

OECD (2014c), *OECD Territorial Reviews: Netherlands 2014*, OECD Publishing, Paris.
DOI: http://dx.doi.org/10.1787/9789264209527-en.

OECD (2014d), *OECD Reviews of Innovation Policy: Netherlands 2014*, OECD Publishing, Paris.
DOI: http://dx.doi.org/10.1787/9789264213159-en.

OECD (2012), *OECD Economic Surveys: Netherlands 2012*, OECD Publishing, Paris.
DOI: http://dx.doi.org/10.1787/eco_surveys-nld-2012-en.

OECD (2008), *OECD Rural Policy Reviews: Netherlands 2008*, OECD Publishing, Paris.
DOI: http://dx.doi.org/10.1787/9789264041974-en.

Annex 1.A1

Recommendations from the OECD Review of Innovation Policy

Maintain supportive framework conditions for innovation and entrepreneurship

- Maintain sustainable public finances as an important requirement for dynamic private and public investment in innovation. In undertaking the required fiscal consolidation, fully take into account the potential negative long-run effects of reducing investments in human capital and in basic and applied research.

- Improve the environment for experimentation by young firms by further improvements in product market regulation, e.g. as regards licensing and permits.

- Pay due attention to the role of a flexible and well-functioning labour market as a precondition for a competitive and entrepreneurial environment, especially for allowing successful young businesses to scale up and find the required set of skills and quality in the workforce.

- Furthermore, contain the cost of dismissals by an appropriate easing of employment protection of open-ended, permanent contracts; this would help facilitate experimentation with business models and foster the reallocation of employment towards the most productive companies.

- Encourage risk financing by continuing to improve the regulatory and legal environment. In particular, consider alleviating restrictions on banks, insurance companies and pension funds for investments in alternative assets such as venture capital.

Improve public governance, steering and co-ordination

- Systematise learning from past experience and tailor governance arrangements to the specific co-ordination problems of each sector. […]

- Refine the strategic vision for the top sectors approach and make a compelling, evidence-based case for sector selection and for the merits of government support. To this end the link between the high aspirations of the approach and existing monitoring tools could be improved by introducing intermediate-level objectives. The continuation of government support could be linked to success in meeting these objectives within specific timeframes.

- Consider extending coverage, or at least transferring valuable experience, to other sectors, in particular those with greater scope to improve the intensity, scope and ambition of their innovation activities. […]

- Extend the monitoring and evaluation framework to take due account of effects outside the top sectors (i.e. full social cost-benefit analyses). In particular, closely monitor its impact on the strong international performance of Dutch fundamental research.

- Institute mechanisms to strengthen the dynamism of the approach in light of societal challenges, emerging technologies and changes in global demand. Among others this may include: Strong representation of smaller and entrepreneurial companies; Use of part of top-sector funding for competitive identification of innovation activities that cut across the top sectors, e.g. multidisciplinary PPPs that could lead to new and valuable combinations of knowledge.

Maintain a world-class human resource base for science, technology and innovation

- Extend the human capital agenda initiative to parts of the economy not covered by the top sectors. [...]

- Strengthen the alignment between top-sector-related skills initiatives and the broader education policy agenda. A close monitoring of the effectiveness of co-ordination in the skills agendas and other measures should enable systematic learning. A key policy task will be to draw broader lessons for national education policy from the top-sector-related initiatives.

- Consider tying a larger portion of block grant allocations to multi-year performance targets. If evaluations show that performance agreements meet their objectives in terms of improving teaching quality, consideration should be given to raising the performance-related component of the block grant allocation, perhaps to as much as 20% of the total.

- Continue efforts to improve teaching quality in higher education institutions, particularly in the UAS. These efforts should ensure the continuing relevance of teaching programmes to the needs of industry, while strengthening generic skills such as innovation management and entrepreneurship. These institution-level efforts should be complemented by a national overview on the provision of teaching programmes so as to avoid "blank spots" in the national coverage of disciplines Further improve the availability of STI skills by stepping up efforts to facilitate

- Further improve the availability of Science, technology and innovation skills be stepping up efforts to facilitate lifelong learning, improve female representation in science and engineering careers, and attract international talent. The Netherlands is already making good progress in each of these areas, but efforts need to continue to offset expected skills shortages caused by an ageing population.

Foster innovation in the business sector

- Raise the intensity, scope and ambition of business innovation. Pay particular attention to the needs of firms that are already innovative (or otherwise knowledge-intensive) but collaborate little with knowledge institutes and conduct little R&D.

- Rebalance the policy mix by complementing the current focus on R&D tax credits with competitive, well-designed direct support instruments, e.g. for joint R&D projects with knowledge institutes, and instruments used in the top sectors approach, such as the MIT.

- Foster stability and minimise the burden imposed on businesses by frequent changes in the policy mix. Predictability could be improved by linking major policy changes to system evaluation cycles agreed upon in advance (e.g. over five-year periods).

Foster critical mass, excellence and relevance of public research

- Continue to nurture the high-quality research performed in the public sector. This means maintaining healthy funding streams for fundamental research, particularly in NWO and KNAW. Avoid an overly strong focus on the top sectors, as this will limit the funds available for new topics and important research areas that are not directly relevant, with the risk that insufficient means are available for new areas and the pursuit of unexpected or risky topics.

- Increase the contribution that business is expected to make to public-private partnerships. The present requirements for business appear to be light, perhaps necessarily so while relationships and arrangements for PPPs still need time to develop. But the rules should be kept under scrutiny, with a view to increasing business commitments. It will be important to ensure that complementarity effects dominate possible crowding out.

- Ensure that valorisation agendas are realistic and take sufficient account of the demand for public research from the business sector. Insofar as the apparent shortcomings in collaboration are due to insufficiently ambitious forms of innovation in parts of the Dutch business sector, there is a danger of too much emphasis on supply-side measures when real bottlenecks persist in the absorptive capacity and behaviour of parts of the business sector. At the same time, it is important for policy to broaden its concept of valorisation, for example by acknowledging the multiple channels through which public research contributes to the economy and society, and to improve its measurement and evaluation accordingly.

- Consider accelerating the development of research capabilities in the UAS, ensuring their close alignment with the main teaching programmes.

- Closely monitor the impacts of funding cuts and the top sectors on the TO2 institutes, bearing in mind that, like their fundamental science counterparts, they require a certain level of stability and continuity, as well as a long-term perspective, for investment in core competences and infrastructure.

- Ensure that the new uniform evaluation arrangements for the TO2 institutes respect the full range of their activities and outputs, as well as the considerable diversity of the institutes. In particular, it will be important to avoid over-reliance on crude indicators, a particular risk when aiming for measurement standardization across institutes

Supporting international knowledge linkages

- Consider developing a co-ordinated national strategy on the international aspects of STI policy. This could be jointly led by the Ministry of Economic Affairs and the Ministry of Education, Culture and Science and would promote co-ordinated action *vis-à-vis* internationalisation agendas more generally.

- Continue to strengthen measures to increase the participation of Dutch SMEs in European programmes for science and innovation. The focus on SMEs and the simplification of rules both in Horizon2020 and the top sectors approach offer an opportunity to target them specifically.

- Continue to promote synergies between the top sectors and the Horizon 2020 agenda while using European programmes to promote science and innovation more broadly, including for sectors and challenges not covered by the top sectors. Establishing a stronger link between top sectors and societal challenges could make an important contribution in this regard. European programmes can also be an opportunity to connect internationally leading research teams active in non top-sector research fields.

Strengthen regional innovation policy and co-ordination between different levels of government

- Continue to engage regional and local authorities in the implementation and definition of regional aspects of top-sector policy (notably support for SMEs and the human capital agenda). This will require the promotion of more active bottom-up consultations between multiple levels of government with the participation of regional and local representatives in steering groups and consultation teams.

- Actively promote links between peripheral but innovative top-sector firms and leading clusters of activities. It is important to ensure that innovative but isolated firms do not lack the opportunity to tap into national and international innovation networks.

- Manage expectations regarding the alignment of the top sectors and regional innovation policy agendas appropriately. Depending on the different regional specialisations, the alignment of top sectors and regional innovation agendas may be more or less appropriate. In addition, the non-alignment of some aspects of regional and local programmes to the top sectors may allow bottom-up initiatives to emerge.

This annex is based on OECD (2014d), *OECD Reviews of Innovation Policy: Netherlands 2014.*

Chapter 2

Overview of the food and agriculture situation in the Netherlands

This chapter describes the overall economic, social and environmental context in which the food and agriculture sector in the Netherlands operates, and the natural resource base upon which it relies. It provides an overview of the general geographical and economic characteristics of the Netherlands; outlines the share of the agri-food complex in the economy; identifies the main structural characteristics of primary agriculture and upstream and downstream industries; provides an overview of the main food and agriculture outputs and markets; and analyses the main trends in agricultural productivity and sustainability. It finally raises a number of issues the agri-food complex is likely to face in the future.

The statistical data for Israel are supplied by and under the responsibility of the relevant Israeli authorities. The use of such data by the OECD is without prejudice to the status of the Golan Heights, East Jerusalem and Israeli settlements in the West Bank under the terms of international law.

General natural and economic context for food and agriculture

With 16.85 million inhabitants in 2013 and a land surface area of 33.7 thousand square kilometres, the Netherlands is the second most densely populated country in the OECD after Korea. About 17% of the entire land area has been reclaimed from the sea or lakes; half of the land in the Netherlands is below sea level, protected by dikes to prevent flooding. The degree of urbanisation is quite high in the Netherlands. Like most OECD countries over the past century, the Netherlands has been undergoing a process of urbanisation. This is partly driven by the structural transformation of the agricultural sector – which is reducing the number of rural inhabitants – as well as by the growth of the service sector. According to the OECD regional typology, urban regions were home to 86% of the national population in 2010 and intermediate ones to the remaining 14%.[1]

Because of proximity to the sea and flat terrain, the country has a mild, maritime climate. Summers are generally warm, and colder, rainy periods, or excessive heat are relatively infrequent. Winters can be cold, windy, with rain and some snow, but extreme cold is rare. Severe weather events occasionally occur when high and low pressure areas meet around the country's territory, which is prone to flooding. A rise in the sea level and more powerful North Sea storms are concerns in the context of climate change.

The natural conditions favour diverse agricultural activities: grassland based and intensive livestock farming, arable farming, open field horticulture and greenhouse horticulture. The geographical position of the country in the heart of North-western Europe favours an important role in international trade.

As for all other economic activities, food and agriculture benefits from policy and economic framework conditions that are conducive to innovation, with good governance and clear rules, competitive and open markets, high quality infrastructure and services, well-educated and skilled workers, and a performant innovation system (Figure 3.1). As a result, the Dutch economy is knowledge-intensive, including in food and agriculture.

The long-standing export-orientation of the economy is reflected in the competitiveness on foreign markets of Dutch agri-food products, mainly livestock products, fruit and vegetables, floriculture and ornamental plants, and processed products (Berkhout et al., 2014). The Netherlands also exports agri-food technologies and know-how, including in the horticultural sector. Export performance in food, agricultural and horticulture is driven by continuous adoption of technological, organisational, product and marketing innovations, which help Dutch farmers and processors maintain a competitive hedge in the EU common market and worldwide.

The EU Common Agricultural Policy (CAP) has influenced the Dutch agri-food sector, in particular the sugar, beef and dairy sectors, but to a lesser extent than in other EU countries given its product orientation, and with contrasting effects depending on the sub-sector. The CAP has traditionally supported crops and extensive livestock systems, while the Netherlands is specialised in intensive livestock systems and horticulture, which are much less protected. While the dairy sector has benefited from high price support, dairy production quotas have constrained the growth of milk production, in a country which achieves very high levels of productivity.

Share of the agricultural complex in the economy

The agri-food complex, which includes the entirety of agricultural and food economic activities – primary production, processing, input manufacturing and distribution –, is knowledge-intensive and technologically highly developed. It accounted for close to 10% of total GDP and employment in 2011 (Figure 2.1.A).

Figure 2.1. Share of the agri-food complex in the economy, 1995, 2005, 2011

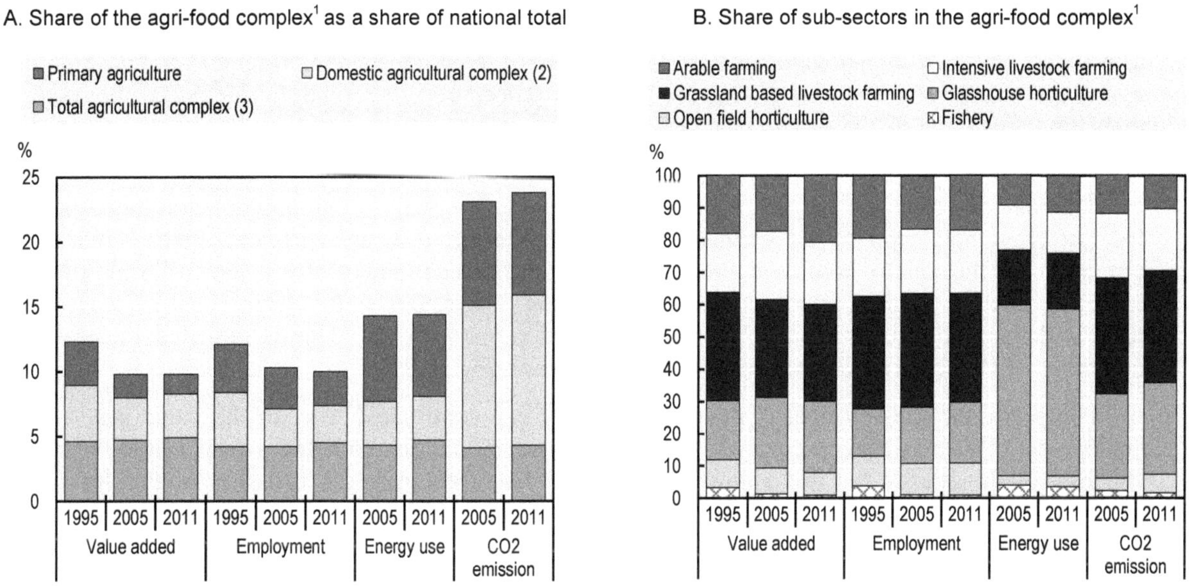

1. The agri-food complex is based on domestic and foreign agricultural raw materials (including gardening, agricultural services, forestry, cocoa, alcohol and tobacco), and includes primary production, processing, input manufacturing and distribution.

2. Based on domestic agricultural resources (excluding cocoa, drinks, tobacco).

3. Based on domestic and imported agricultural resources.

Source: Figures for primary agriculture come from OECD and Eurostat databases. Figures for the agricultural complex come from Berkhout, P., H.J. Silvis, and I. Terluin (eds.) (2014), *Agricultural Economic Report 2014 of the Netherlands: Summary*, http://www.landbouweconomischbericht.nl/download/summary-2013-pdf.html.

StatLink http://dx.doi.org/10.1787/888933250744

Just over half of the activities in the agri-food complex are to a greater or lesser extent directly related to the primary Dutch agriculture and horticulture, and accounts for 15% of the value-added and about 15% of the employment of the agri-food complex (Figure 2.1.B, Table 2.A1.1). The remainder relates to the supply, processing and distribution of (international) raw materials and to gardening and forestry. Some downstream segments operate entirely to add value to imported raw products. For example, the Netherlands does not produce cocoa, but it is the world's leading producer of cocoa powder and cocoa butter.

Between 2005 and 2011 the contribution of the agricultural complex to national value-added stabilised, whereas its contribution to national employment declined, as lower employment in primary agriculture was not compensated by an increase in employment in downstream industries (Table 2.A1.1). The dependency of the agri-food complex on activities related to the processing, delivery and distribution of imported raw agricultural materials has been growing. Exports generate about three-quarters of the value-added and employment of the agri-food complex (Berkhout et al., 2014).

The agri-food complex contributes significantly to the Dutch export revenues. Whereas the Dutch share of total world trade is estimated at 3.3%, the export of the agro-food sector reached 7.8% of world total agro-food exports in 2010, and this has been on the 6-8% level for the last four decades (Snijders and Jacobs, 2013). Agro-food trade (food products, beverages and tobacco) also accounts for a significant share of total Dutch trade (13% for exports and 9% for imports in 2013) and makes between a third and half the national trade surplus depending on the year.

Figure 2.1.B distinguishes six sub-complexes:[2] arable farming, open field horticulture, glasshouse horticulture, grassland-based livestock farming, intensive livestock farming and fishery. Although there are similarities between the sectors and chains in the agricultural complex, there are also major differences, which lie in, among other things, the nature of the products, the origin and composition of inputs, the structure of the chains (e.g. the share of processing), the financial situation, the export orientation, but also in the environmental externalities generated by activities along the chain, in particular in the primary sector, and related government interventions.

The majority of Dutch farms function as part of vertical systems, with coordination mechanisms that span from input supplies to consumer level. Another feature of Dutch agri-food chains is that they extend beyond the country borders and rely on external markets as sources of supplies for local primary agriculture and processing. Imported feeds are supplied to livestock farms, which convert them into meat or dairy products, which are in turn directed to export markets. This is reflected in the high integration of Dutch agri-food sector in global value chains (Figure 2.3).

For years, the grassland-based livestock complex has contributed most to the value added and employment of the agricultural complex. The greenhouse horticulture complex is responsible for more than half of the energy use and almost one-third of the greenhouse gas emission of the agri-food complex. The livestock sector also accounts for about a third of the greenhouse gas (GHG) emission of the agri-food complex.

Structural characteristics of farms

The total acreage of cultivated land in use by registered agricultural and horticultural holdings amounts to 1.85 million ha (Table 2.1). Two-thirds of agricultural land is used for grass and fodder (Figure 2.2).

Table 2.1. Selected indicators of the farming sector, 2000, 2013

Indicator	2000	2013	% change
Area farmland (1 000 ha)	1 975	1 848	-6.7
Employment (1 000 AWU)[1]	n.a.	160.5	-30%[3]
Total number of holdings[2]	97 389	67 481	-30.7
- Arable farms	14 799	12 142	-18.0
- Glasshouse horticulture and mushroom holdings	8 804	3 794	-56.9
- Dairy farms	23 280	17 001	-27.0
- Other grassland based livestock farms	20 208	17 757	-12.1
- Intensive livestock farms	12 058	6 744	-44.1

n.a.: not available because of break in series. AWU: Annual Work Unit.

1. Agriculture, forestry and fisheries (NACE Rev2). About 58% is family labour (not paid).

2. In 2013 the number of holdings declined by 1.9% compared to 2012.

3. Based on number of persons.

Source: Statistics Netherlands and LEI, http://www.agrimatie.nl/ .

StatLink http://dx.doi.org/10.1787/888933251295

Figure 2.2. Agricultural land use, 2013

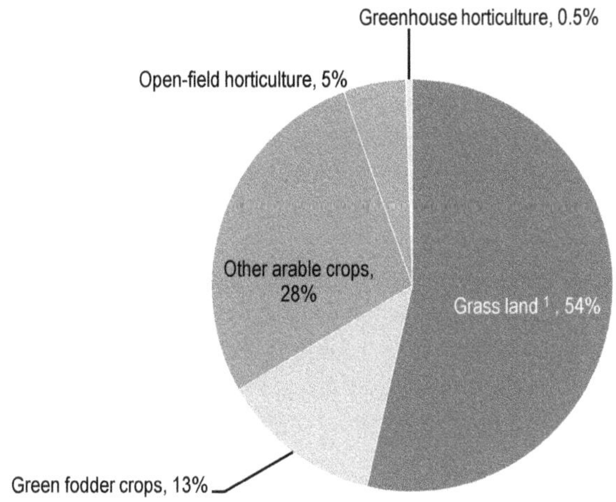

1. Permanent, temporary and natural grassland.
Source: Statistics Netherlands.

StatLink http://dx.doi.org/10.1787/888933250758

Decreasing employment and number of holdings in agriculture and horticulture

The employment provided by the primary agricultural and horticultural sector, expressed in terms of employees working on a regular basis, has declined by 30% since 2000, from 281 000 to 198 000 in 2012 (this corresponds with 160 000 annual work units). The number of agricultural and horticultural holdings is decreasing at a rate of 2 to 3% per year (Table 2.1). The decrease in the number of holdings varies between the sectors, ranging from 18% for arable farms between 2000 and 2012 to 57% for greenhouse horticulture. The greatest contraction is observed in the number of mixed farms (almost 60%), a decline which confirms that the structural developments in the agricultural and horticultural sector are characterised by specialisation together with increases in scale.

Forces behind the continuing decline in the number of holdings are the age distribution of the holders, the availability of a successor and technological developments including labour-saving innovation developments. The fairly large decline in the number of non-land-based holdings since the turn of the century is due to factors including environmental and animal welfare policies (such as mandatory investments to comply to new regulations) and market developments.

The decline in the number of holdings is primarily due to the next generation deciding not to take over the family business in view of the moderate income prospects compared to other employment alternatives. Forced termination in the form of a bankruptcy is rare. More than 800 agricultural and horticultural holdings have been declared bankrupt since 2000, equivalent to almost 3% of the total decline in the number of holdings. The sharp increase in the number of bankruptcies in the overall Dutch economy since the beginning of the economic crisis in 2008 is also reflected in the agricultural and horticultural sector. By far the most bankruptcies were in the plant sectors (approximately 90% in the past four years). Most of these are greenhouse horticulture holdings.

Increasing size of agricultural holdings

With the decline in the number of farms, farm size increased dramatically. The mid-point farm size number, which splits the distribution of animal numbers equally, show that between 2000 and 2013, the mid-point weighted median herd size has increased from 60 to 90 dairy cows in specialised dairy farms, 1 000 to 3 000 pigs in specialised pig farms, and 60 000 to 120 000 birds in specialised poultry farms, while the mid-point area of greenhouse horticulture farms increased from 2 to 7 ha.

With the increasing size of holdings, the workforce per holding has gradually increased from 1.9 employees at the beginning of the 1990s to 2.3 employees per holding in recent years. Ownership of virtually all agricultural and horticultural holdings stays within the family and holdings are transferred between generations. Farm acquisition faces a number of difficulties including increasing size of holdings and land values, more business-like family relationships regarding investment, and the stringent requirements imposed on bank loans. The increasing size of holdings and higher land values confronts successors with the need to finance larger takeover sums. Dairy and arable farms have invested a large amount of capital in their land, and their equity and solvency tends to rise further due to increases in land prices. When taking over an existing operation, the successor often relies on loan arrangement within the family, which explains why the share of family loans in the loan capital is much larger in the land-based sectors as compared to the average for the entire sector. These family loans play an important role in meeting the conditions attached to taking over holdings in these sectors. Because the amounts of money in taking over an existing operation are substantial, holdings taken over in recent years often have the worst equity position,although they are very large and modern.

Income trends

Farm income varies greatly, in part due to differences in the size and structure of the farm holdings. Larger holdings are included in both the groups with the highest and the lowest income. Incomes can also fluctuate widely from year to year. Since 2005, the real income of factors per annual agricultural work unit (AWU) increased by 56% in the Netherlands, compared to 19% in the EU15 and 34% in the EU28 (EUROSTAT, 2014). On average, income from outside the holding accounts for almost one-third of the holder's income (Berkhout et al., 2014). Between 2007 and 2013, income from on-farm non-agricultural activities using farm inputs increased by more than 60% but remains limited. Recreation is the most important activity, followed by on-farm sales of agricultural products. Agricultural childcare and care farming have shown the most growth in turnover in recent years (EUROSTAT, 2014).

Compared to other sectors, the agricultural and horticultural sector records a low yield on net assets, a yield that is largely based on the revaluation of land (Berkhout et al., 2014). On average, just 47% of the calculated costs of own labour and capital are reimbursed from the holding's income.

Structural characteristics of upstream and downstream industries

Supply industries

The supply industries of the agricultural sector are heterogeneous, and include the production of machinery, fertiliser, plant protection products, construction and compound feed.

The Netherlands is an important producer and exporter of fertilisers, in particular nitrogenous fertilisers. In 2011, the Netherlands produced 1.5 million tonnes of nitrogenous fertilisers (N) and 122 500 tonnes of phosphate fertilisers (P_2O_5). More than 90% of Dutch production is exported. The sector is an important supplier of the primary agricultural and horticultural sector and is closely related to the livestock farming sector. The fertiliser manufacturers (ICL Fertilizers Europe, OCI Nitrogen, Yara and Rosier Nederland) in the Netherlands all have a foreign parent company.

The greenhouse construction industry is, together with the greenhouse installation and technical equipment industry, closely related to the greenhouse horticulture sector. Many greenhouse construction companies are also active abroad: about half of their turnover from outside the Netherlands is generated in Western Europe. Many Dutch greenhouse construction companies are also active in the Russian Federation, Turkey, Mexico, East Africa and the Far East, where they have leading positions in their respective markets. Estimates indicate that about 80% of all greenhouses in use outside Europe are of Dutch origins. Greenhouse construction in the Netherlands is confronted with difficult conditions due to the continuing poor results recorded by the Dutch greenhouse horticulture sector in recent years, which have compelled many growers to cut back their investments to a low level.

Food processing industries

Food and beverage industries account for 21% of industry turn-over, and 9% of firms (Table 2.2). The average size of companies is larger than the industry average, with a largest average size in the edible oils and fats industry, the tobacco industry, the animal feed industry and the dairy processing industry, while smaller firms are found in meat processing, fish processing, bakery and cake industry, and bread and pasta industry. Dairy and meat processing account together for 28% of the turnover of food and beverage industries, and edible oils and fats, and animal feed industries, for about 12% of the total each.

Animal feed

The largest segment of the agricultural supply industry is the animal feed industry. It is closely related to the dairy industry and to the abattoirs and the meat-processing industry, as the feed industry is a crucial supplier of the Dutch livestock farming sector. The Dutch animal feed industry's main products are pig feeds (40%), followed by poultry feeds (27%), and cattle feeds (24%). Mergers and takeovers have resulted in the formation of a number of large Dutch multinationals that rank among the top European businesses, such as Forfarmers, Agrifirm and De Heus Voeders.

Table 2.2. Key figures of the food and beverages industry in the Netherlands, 2012

	Number of enterprises	Employment (x 1 000)	Turnover[1] (EUR Million)	% share in turnover of all food and beverage industry
Total industry	**53 930**	**887.4**	**316 566**	
Food and beverages industry	4 780	155.4	67 316	100.0
Of which,[2]				
Fruit and vegetable processing industry	150	13.7	4 797	7.1
- Potato products industry	35	5.2	2 124	3.2
Edible oils and fats industry	40	3.1	8 292	12.3
Flour industry	110	3.6	2 326	3.5
Bread and pasta industry	2 515	47.4	4 586	6.8
Bakery and cake industry	150	7.4	1 426	2.1
Animal feed industry	175	7.7	8 069	12.0
Dairy (processing)	300	13.2	9 235	13.7
Slaughterhouses and meat processing industry	550	24.1	9 845	14.6
- Slaughterhouses (excl. Poultry)	325	9.4	4 611	6.8
- Slaughterhouses poultry	40	6.3	2 669	4.0
- Meat processing	185	8.4	2 566	3.8
Fish processing industry	130	4.0	827	1.2
Cocoa and chocolate confectionery industry	190	8.3	4 331	6.4
Beverages industry	215	7.8	4 835	7.2
- Soft drink industry	20	2.4	1 468	2.2
Tobacco industry	20	3.0	2 526	3.8

1. Returns (excluding VAT) from the sales of goods and services to third parties.
2. The list of sub-sectors is not comprehensive.

Source: Statistics Netherlands. http://www.cbs.nl/en-GB/menu/home/default.htm.

StatLink http://dx.doi.org/10.1787/888933251304

A special role for co-operatives

Agricultural co-operatives play a special role in upstream and downstream industries (Table 2.3). There has been a strong merger process between co-operatives leading to only a few or even one co-operative per sector. Federated co-operatives have disappeared, except in banking where Rabobank is an important player. Investor-owned firms have left several sectors like sugar and starch potato, while in other sectors (like slaughterhouses) co-operatives have been less successful. Several co-operatives are international or even transnational (like the dairy company FrieslandCampina).

Table 2.3. Market shares of co-operatives in the Netherlands, 2000-2010

	Number of agricultural co-operatives		Market share (%)		Number of farmer members in the Netherlands	
	2000	2010	2000	2010	2000	2010
Farm inputs	25	15	n.a.	n.a.	50 000	35 000
Animal feed	20	13	53	55	n.a.	28 000
Pig breeding	1	1	n.a.	85	n.a.	2 300
Cattle breeding	1	1	90	80-90	34 750	18 000
Sugar	3	2	63	100	13 700	9 940
Dairy (processing)	5	5	83	86	21 600	15 200
Pig meat	3	0	34	0	10 000	0
Fruit & vegetables	15	19	71	95	9 000	4 500
Potato starch	1	1	100	100	4 800	1 600
Seed and ware potatoes	7	6	n.a.	n.a.	3 900	1 500
Mushrooms	2	3	>50	>80	470	200
Flowers	6	3	95	95	9 400	5 300

n.a.: not available.

Source: Bijman, J. et al. (2012), *Support for Farmers' Co-operatives; Country Report,* http://edepot.wur.nl/244818.

StatLink http://dx.doi.org/10.1787/888933251319

The farmers' co-operatives in the Netherlands are competitive in the "first transformation" stage, and some of them also in marketing branded food products. They organise product supply, which facilitates access to the domestic and international markets. The traditional farmer-owned co-operative in the food chain was very close to the producer and rather distant from the consumer. This has been changing significantly over the last two decades. Co-operatives have since adopted the market strategies of (international) Investor-Oriented Firms in the food industry. Those firms have had a stronger focus on developing and marketing branded products. Dutch co-operatives have substantially increased their effort in product innovation and marketing branded products since the 1980s.

As co-operatives evolved and diversified, so did their strategies. Co-operatives traditionally followed a cost-leadership strategy, increasing the efficiency of their processing and sales operations. For instance, sugar co-operative Cosun is known in Europe as a low-cost producer of sugar. Members have always urged their co-operative to keep operational costs as low as possible. Co-operatives were not able to influence the price, as they were price takers in very competitive markets, or prices were determined by the Common Agricultural Policy (CAP). This cost-leadership strategy has led to many mergers among co-operatives when technological developments raised the minimum efficient scale of operation.

Marketing co-operatives (in the dairy, sugar, and starch potato industry) have developed new consumer and industrial products based on the ingredients of the commodity supplied by their members (milk, sugar beet, starch potato). Next, co-operatives producing final consumer goods have developed own brands to strengthen the competitive position, both horizontally towards other food companies and vertically towards large food retailers (supermarkets).

Fruit and vegetables

The fresh produce industry has shown contradictory developments over the last 20 years. On the one hand many crop or variety specific producer organisations (POs) have been established. The importance of product innovation and market segmentation was one of the drivers of this development. These POs were either an alternative for the traditional auction co-operatives, or they were complementary organisations as they were established by co-operative members who wanted to get more attention for their specific product. On the other hand there is an ongoing trend of collaboration and mergers among POs in order to benefit from economies of scale and to become (or remain) an attractive partner for the food retail sector.

Dominance of supermarkets

A major characteristic of the food chain is the dominance of supermarkets (Table 2.4). Most food is sold through supermarkets. The dominance of supermarkets in food chains has affected the supply chains in several ways. First, supermarkets put special emphasis on quality and food safety assurance, and product attributes such as sustainability and animal welfare. Second, suppliers to supermarkets need to offer the full assortment of products all year-round. Supermarkets like to purchase from a small group of preferred suppliers. This has contributed to the concentration of processing companies in the food supply chains.

Table 2.4. Key figures of main retailers, 2013

Company	Store	Turnover (Million EUR)	Number of stores	Market share (%)
Ahold	Albert Heijn (AH)	11 054[1]	907 (AH)	34.0
Jumbo Supermarkten	Jumbo Groep	7 092	658	20.1
Lidl Nederland	Lidl		379	9.0
Aldi Nederland	Aldi		503	7.4
Sperwer Groep	PLUS	1 970	254	5.8

1. Total turnover in the Netherlands, including Gall & Gall.

Source: Berkhout, P., et al. (eds.) (2014), *Landbouw-Economisch Bericht 2014*, http://edepot.wur.nl/306953.

StatLink http://dx.doi.org/10.1787/888933251320

Although the share of online sales in the total turnover of supermarkets is still small at 1% of total sales, it is expected to expand to 15-20% (Berkhout et al., 2014). Retailer Ahold prepares itself for an era in which the supply and purchase increasingly takes place through the Internet. With the integrated webshop ah.nl and the acquisition of bol.com it has become a leading company in online shopping.

Changes in the structure and drivers of the supply chain

The food economy refers to the entire supply chain from farm input industries and research institutes developing food innovations to the final retail point and food consumer. Traditional supply chains comprised of a linear set of relationships, from producers, through sellers and finally to consumers. The modern concept of net chains (a combination of networks and chains) is a web or network of relationships centred on consumers, where other stakeholders (including NGOs) are involved explicitly, and information exchange is co-ordinated by network members.

Technological innovation used to be oriented towards agriculture and food processing, aimed at improving efficiency. Now, the development of networks is leading decisions about what to produce, where to sell it, and at what price. Power in the net chain has shifted from producers to retailers, which reflect consumer demand. There is more urgency to make distinct products preferably with brand names so that substitution with other suppliers becomes more difficult.

Agricultural trade

A substantial part of the operations in the agricultural production chain is related to international trade. Around two-thirds of the total agricultural complex added value and employment is dependent on exports (Table 2.5). According to OECD-WTO estimates of trade in value-added,[3] the Netherlands is a large exporter of agro-food value-added relative to its agro-food GDP (Figure 2.3). Trade in value added also shows that the share of foreign inputs in Dutch exports of agriculture and food products is high compared to other OECD countries and emerging economies (around 20% for agriculture and 40% for food), and that the participation of the Netherlands in global value chains is one of the highest in terms of the use of domestic products in food exports (Miroudot, 2014).

Figure 2.3. Trade in agriculture and food value-added

Agro-food exports and imports of value-added as a percentage of agro-food adjusted GDP (value-added), 2009

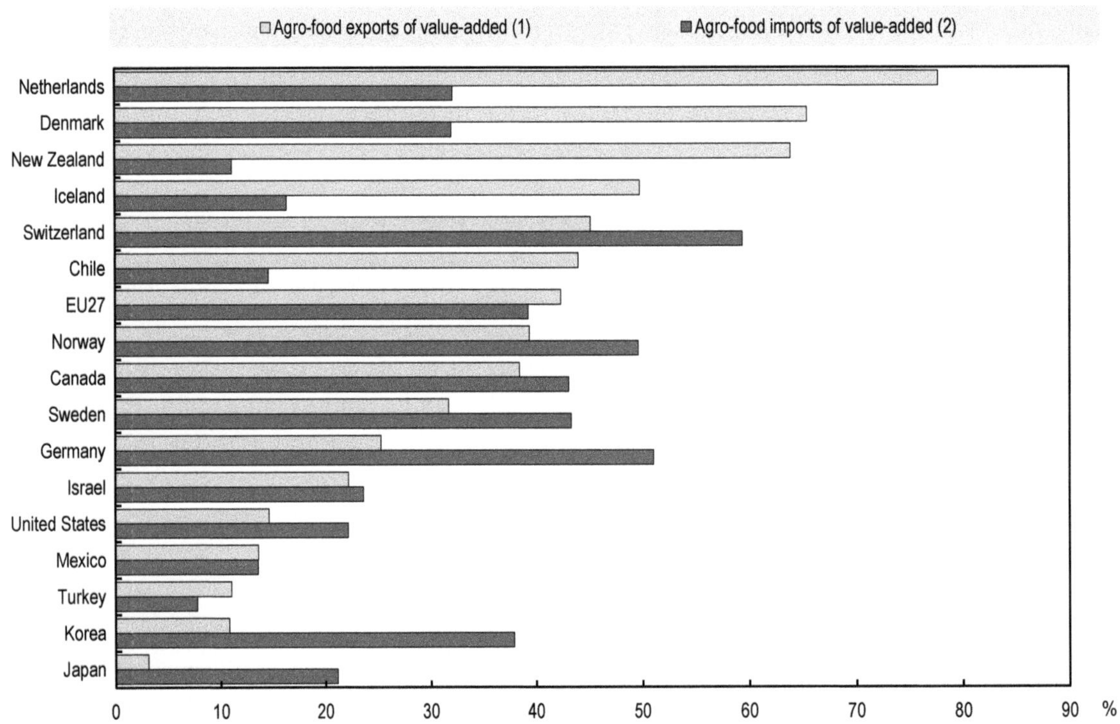

1. Value-Added embodied in Foreign Final Domestic Demand shows how industries export value both through direct final exports and via indirect exports of intermediates through other countries to foreign final consumers. They reflect how industries (upstream in a value-chain) are connected to consumers in other countries, even where no direct trade relationship exists. The indicator illustrates therefore the full upstream impact of final demand in foreign markets to domestic output. It can most readily be interpreted as "exports of value-added".

2. Foreign Value-Added embodied in Final Domestic Demand shows how industries abroad (upstream in a value-chain) are connected to consumers at home, even where no direct trade relationship exists. It can most readily be interpreted as "imports of value-added".

Source: OECD-WTO Trade in Value-Added Database, 2013.
http://www.oecd.org/industry/ind/measuringtradeinvalue-addedanoecd-wtojointinitiative.htm.

StatLink http://dx.doi.org/10.1787/888933250761

The vast majority of Dutch agro-food trade takes place within the European Union. Around 80% of Dutch agricultural exports go to other EU member states, while about 60% of agro-food imports come from the European Union. Within the European Union, the Netherlands' nearest neighbours are its most important trade partners. Approximately one-third of the Netherlands' intra-EU agro-food trade is carried out with Germany; and trade with Belgium, Luxembourg, France and the United Kingdom combined amounts to about 40%. This is similar to the percentages for Dutch intra-EU trade as a whole.

The most important import products from the European Union are dairy, meat, grain and grain preparations. The dairy imports primarily consist of unprocessed milk, whey and skimmed milk powder. These products are used as raw materials in the food and alcohol industries. The agricultural imports from third countries (non-EU countries) consist of margarine, fats and oils, raw materials for animal feed, fruit (including tropical fruit) and coffee and cocoa beans.

Meat, ornamental plants and dairy are the three largest agricultural exports to the European Union. Agro-food exports to third countries primarily consist of processed products such as beverages, dairy products, grain preparations and ornamental plants.

The agro-food sector reaches high world export shares, with notable spikes. Table 2.5 lists the 20 top products at 4-digit level, ranked according to their world export share. Each product has a positive trade balance and takes up more than 0.1% of total Dutch exports, i.e. about EUR 250 million. Dutch exports for each listed product account for a significant share of world trade, ranging from 13% for veal or cheese to 76% for flower bulbs.

Table 2.5. Agro-food top 20 products in export share, 2010

Nr	HS-code	Description	Share in (%)		Value (billion USD)	
			World agro-food exports	Dutch total exports	Exports	Net exports
1	0601	Flower bulbs	76.1	0.3	1.1	1.0
2	0603	Cut flowers	50.5	0.9	3.7	3.1
3	0602	Plants	48.6	0.8	3.3	2.8
4	1805	Cocoa powder	39.7	0.2	0.9	0.8
5	0707	Cucumbers	38.0	0.1	0.5	0.4
6	1803	Cocoa pasta	37.4	0.2	0.7	0.4
7	1804	Cocoa butter	35.4	0.3	1.4	1.0
8	0103	Pigs	32.0	0.3	1.2	1.0
9	0407	Eggs	29.6	0.2	0.8	0.6
10	0702	Tomatoes	27.5	0.4	1.7	1.4
11	1209	Seeds	24.4	0.3	1.2	0.8
12	2004	Other processed vegetables	24.0	0.3	1.3	1.0
13	0701	Potatoes	23.8	0.2	0.8	0.5
14	2402	Cigars and cigarettes	20.7	0.8	3.3	2.8
15	2203	Beer	17.5	0.4	1.8	1.6
16	0210	Pork	16.6	0.1	0.6	0.3
17	0703	Onions	14.2	0.2	0.8	0.6
18	2309	Animal feed	14.0	0.6	2.5	1.8
19	0406	Cheese	13.5	0.8	3.2	2.4
20	0201	Veal	13.2	0.5	2.2	1.0
		Total top 20	**22.8**	**7.9**	**33.0**	**25.3**

Source: Snijders, H. and D. Jacobs (2013), *Clusters en niches; de specialisatie van de Nederlandse economie*, http://www.wrr.nl/fileadmin/nl/publicaties/PDF-webpublicaties/76_Clusters_en_niches._De_specialisatie_van_de_Nederlandse_economie.pdf.

StatLink http://dx.doi.org/10.1787/888933251338

Trends in productivity and sustainability

Productivity

In comparison with other countries, the Netherlands achieves a very high agricultural production value per hectare of cultivated land. This is partly explained by the production package with a relatively large share of horticulture and livestock farming. High productivity per hectare, animal and worker explains the export surplus of domestic agricultural products. Dutch agriculture achieves one of the highest yield performance per ha of cereals among EU member states and one of the highest milk yields per dairy cow (OECD, 2011). In 2011, with EUR 85 000 per labour unit, the labour productivity of the arable complex was the highest of all sub-complexes, followed by the glasshouse horticultural complex (Van Leeuwen et al., 2014).

Total factor productivity (TFP) has increased by 2.6% per year on average over the period 1961-2011. Until 1990, TFP growth was over 2% per year and mainly driven by increases in production. In the last two decades, it has been driven by a reduction in input use due to policies that restricted the use of certain inputs (e.g. pesticides policy, Nitrates Directive) (Figure 2.4). After a decade of relatively slow TFP growth in the 1990s due to the stagnation of production, the Netherlands has regained a leading position in the first decade of the 21st century as by optimising input use, it has often been possible to use fewer inputs, sometimes at a lower cost, to obtain the same output (Figure 2.5).

Both structural change and the adoption of various types of innovations on the farm have enabled sustained TFP growth. Box 2.1 outlines the link between productivity and on-farm innovation in Dutch dairy farms over the last decade.

Figure 2.4. Trends in the Total Factor Productivity of Dutch primary agriculture, 1961-2011

Index 1992=100

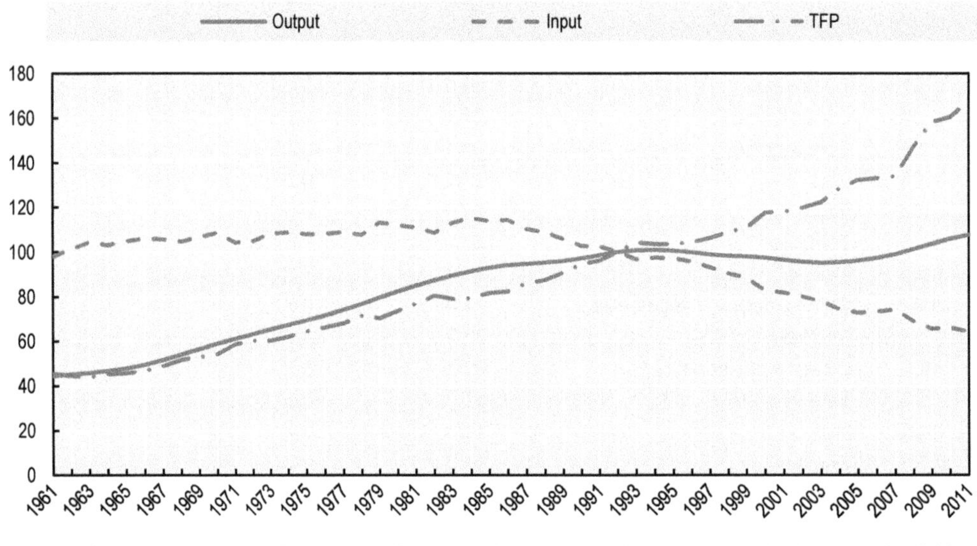

Source: USDA Economic Research Service Agricultural Productivity database. Available at: www.ers.usda.gov/data-products/international-agricultural-productivity/documentation-and-methods.aspx#excel.

StatLink http://dx.doi.org/10.1787/888933250774

Figure 2.5. Total Factor Productivity growth in primary agriculture, international comparison, 1991-2000, 2001-10

Annual percentage growth rate by decade

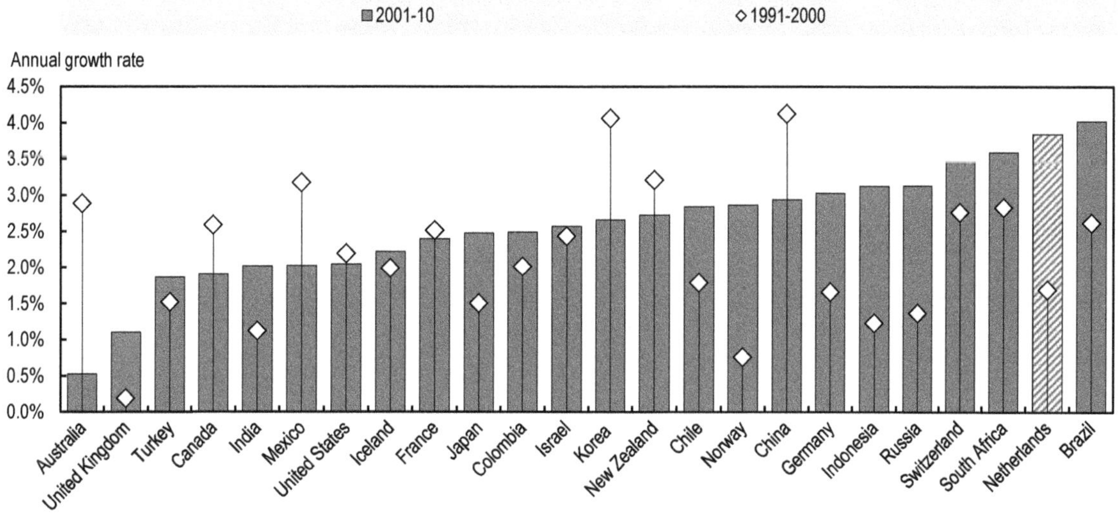

Source: USDA Economic Research Service Agricultural Productivity database. Available at: www.ers.usda.gov/data-products/international-agricultural-productivity/documentation-and-methods.aspx#excel.

StatLink http://dx.doi.org/10.1787/888933250783

Box 2.1. Dynamics of dairy farm productivity growth in the Netherlands

The EU milk quota system, which has been phased out over 2008-15, has been affecting the productivity growth dynamics of the Dutch dairy farm sector (OECD, 2015). In the Netherlands, milk deliveries continued to exceed the national quota marginally, resulting in the payments of surplus levies in the last ten years, except for 2005/06 and 2012/13. The quota price in the Netherlands remained relatively high even after the announcement of the milk quota reform. The price of milk quotas fell from EUR 40 per kg of fat to EUR 18 in one year from July 2006 and further decreased to nearly EUR 10 in 2013. Due to the quota system in place, milk output has been fairly static in the Netherlands until the quota started to be relaxed after the European Union decided to increase milk quotas by 2% in 2008 and increasing milk quotas by 1% annually over five years until their abolition in 2015.

The specialist dairy farm sector has recorded an average annual TFP growth of 1.3% over the period 2001-12, with different growth paths during the periods 2001-07 and 2007-12 (Figure 2.6). The quota system has been a constraint to expanding production in Dutch dairy farms. As a result, farmers have strived for efficiency by reducing costs. Total input use declined on average by 1.7% annually due to a decrease in labour, and material and service inputs. Labour input decreased by 3.2% in 2001-07 and 1.6% in 2007-12, resulting in a 22% decline in labour input over the whole period. It led to a continuous growth in labour productivity of 3.3% per year on average. The productivity improvement during 2001-07 was also driven by a reduction of material and service inputs, which declined by 2.8% and 1.9% annually respectively. In contrast, after the announcement of the milk quota reform in 2007, the growth in output outpaced that in input. This policy change induced a capital investment, which increased the average annual growth rate of capital input from 1.2% to 4.7% after 2007. Although the literature suggests that milk quotas tend to slow down structural change and maintain inefficiencies, the evolution of TFP in the Netherlands implies that the productivity improvement could occur under a milk quota regime with a well-functioning quota market, as milk quotas in the Netherlands could be traded freely.

The TFP growth of the Dutch dairy farm sector in 2001-12 is decomposed into three productivity growth components: within-farm productivity growth, resource reallocation effect and entry-exit effect (Figure 2.7). The productivity improvement at the farm level (within-farm productivity growth) is the main driver of the productivity growth of the Dutch dairy farm sector, accounting on average for more than 70% of annual productivity growth in 2001-12. The contribution of farm-level productivity improvement is found particularly higher in the Netherlands than in two other countries studied in the report: Estonia and the United Kingdom. The contribution of farm-level productivity growth is larger in 2001-07 than in 2008-12. This indicates a trend of strong on-farm innovation in the sector in 2001-07 to produce a stable amount of output with less inputs such as the adoption of labour-saving technology and more efficient management of material and service inputs at the farm level. The regression analysis also shows the positive impacts of the milk robot and milk parlor technology on the farm-level TFP in the Netherlands.

Box 2.1. Dynamics of dairy farm productivity growth in the Netherlands *(cont.)*

The resource reallocation effect contributed positively to productivity growth, accounting for 29% of annual productivity growth in 2001-12 on average, i.e. resources were allocated to more productive farms. In the Netherlands, the average herd size of the largest 25% of dairy farms increased from 107 to 153 dairy cows in 2003-12, with a share in milk production increasing from 42% to 46%. The slightly negative average contribution of farm entry and exit reflects the relative lower rate of farm exit in the Dutch dairy farming sector, but the average productivity contribution of farm entry-exit increases to 18% after 2007 presumably due to exits of unproductive dairy farms. As a result, productivity difference between farms has decreased overtime together with the diffusion of existing technology across farms.

Figure 2.6. Evolution of TFP, output and input indices of Dutch dairy farm sector

2003 = 100

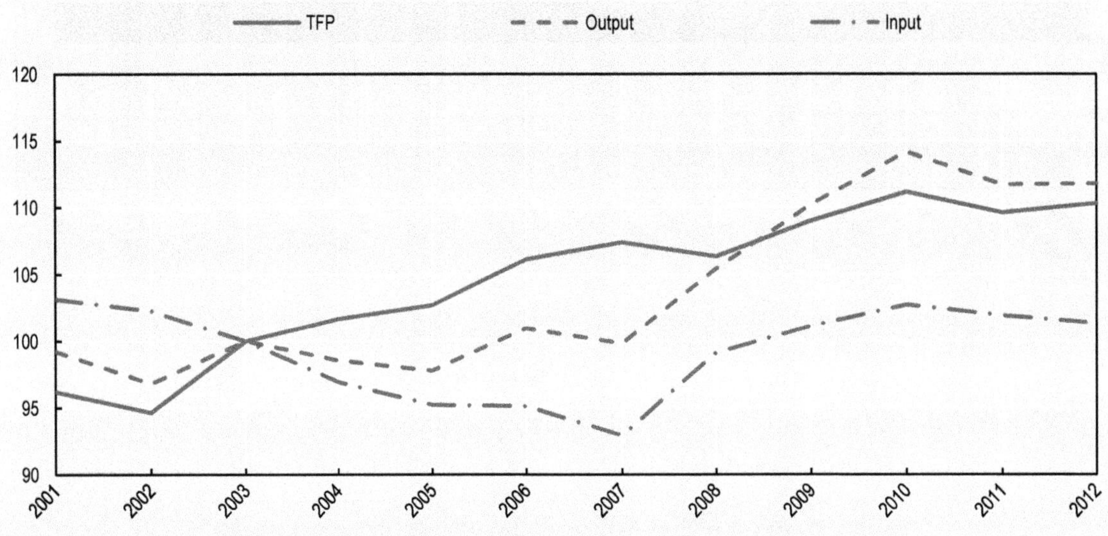

StatLink http://dx.doi.org/10.1787/888933250797

Figure 2.7. Decomposition of TFP growth in the Dutch dairy farm sector

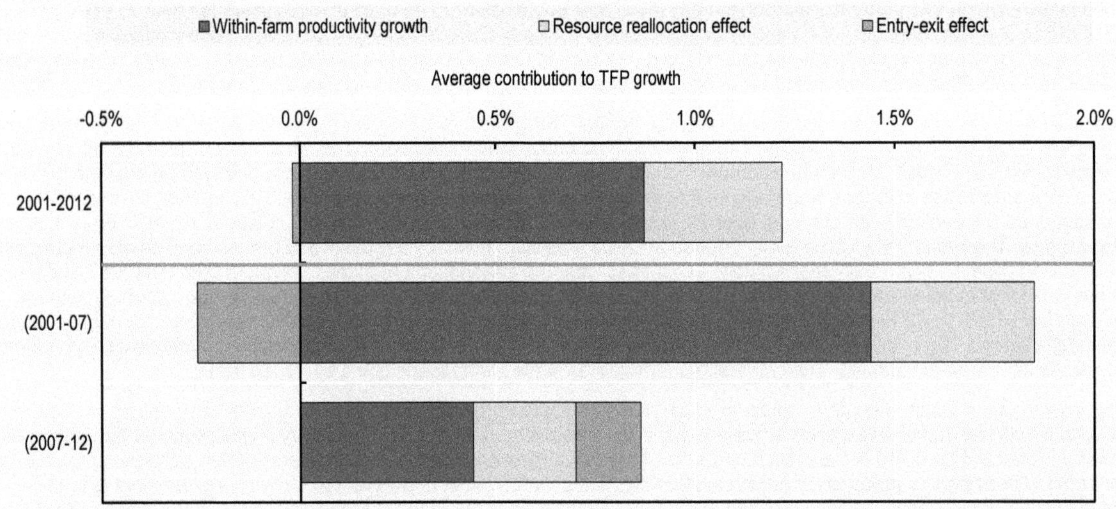

StatLink http://dx.doi.org/10.1787/888933250800

Source: Kimura and Sauer (2015), "Dynamics of Dairy Farm Productivity Growth: Cross-country Comparison".

Sustainability and environmental performance

Connected to the high productivity, the use of inputs and the emissions per hectare are at a high level in the Netherlands. Such intensive pattern of agriculture in the country is source of a set of environmental pressures, including: 1) emissions of nutrients in air, soil and water bodies and ammonia emissions linked to intensive use of fertilisers and livestock activities; 2) risks of pollution associated with the use of pesticides; 3) energy consumption and associated GHG linked to horticulture emissions; 4) GHG emissions (CH4 and N2O) from livestock farming; and 5) pressures on biodiversity. Notwithstanding the progress that the Netherlands made in the 2000s to address the negative impact of environmental pressures, particularly in comparison with the OECD average, several agri-environmental indicators are still above the OECD average and further efforts are required in order to counteract the negative consequences of these practices. Over the past two decades, the Dutch agriculture production volume has almost stabilised while most pressures on the environment from agriculture have been decreasing, as shown by Figure 2.8.

Figure 2.8. Development of environmental indicators for agriculture, Netherlands, OECD and EU15 and the Netherlands, 1998-2000 to 2008-10

Average annual percentage change between 1998-2000 and 2008-10[1]

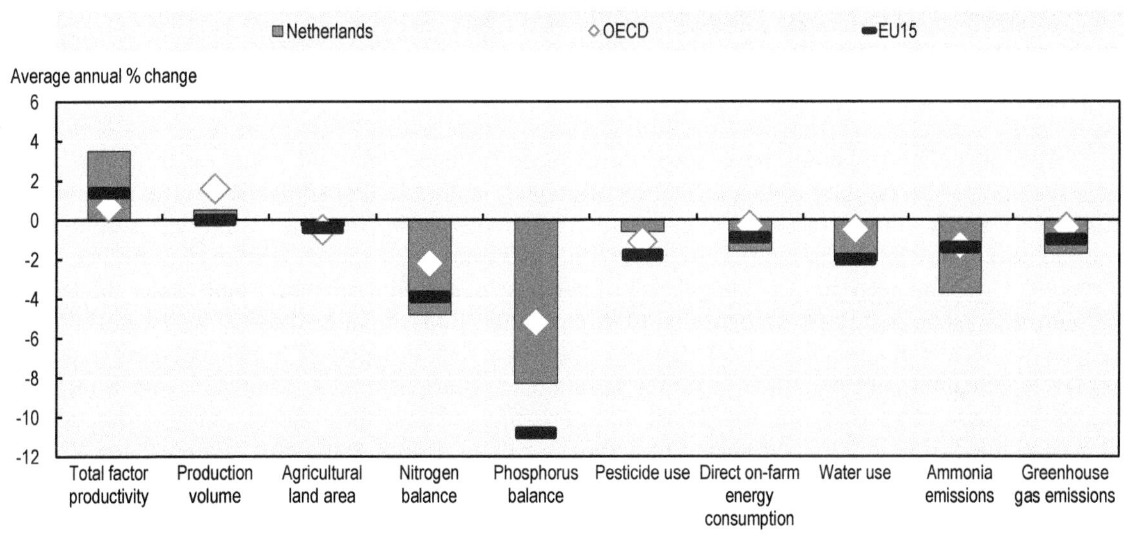

1. 2001-10 for total factor productivity.

Source: OECD (2013), *OECD Compendium of Agri-environmental Indicators*, http://dx.doi.org/10.1787/9789264181151-en. USDA Economic Research Service Agricultural Productivity database, for total factor productivity growth, available at: www.ers.usda.gov/data-products/international-agricultural-productivity/documentation-and-methods.aspx#excel.

StatLink http://dx.doi.org/10.1787/888933250814

Nutrient surplus

Among OECD countries, the Netherlands has the largest surpluses of nitrogen and phosphorous per hectare (with the exception of South Korea). In 2012, the Dutch Gross Nitrogen Balance (GNB) per hectare of agricultural land area was estimated at 163 kg of nitrogen per hectare of agricultural land area (kg N/ha), compared to an EU28 average of 47 kg N/ha.[4] Figure 2.9 provides comparisons with other EU member states such as Belgium, Denmark, Ireland, Germany, France and the United Kingdom. Such a high level is mostly due to land-intensive livestock production associated with high levels of feedstuff import.

Figure 2.9. Gross nitrogen balance in a set of European Union Member States, 1990-2010

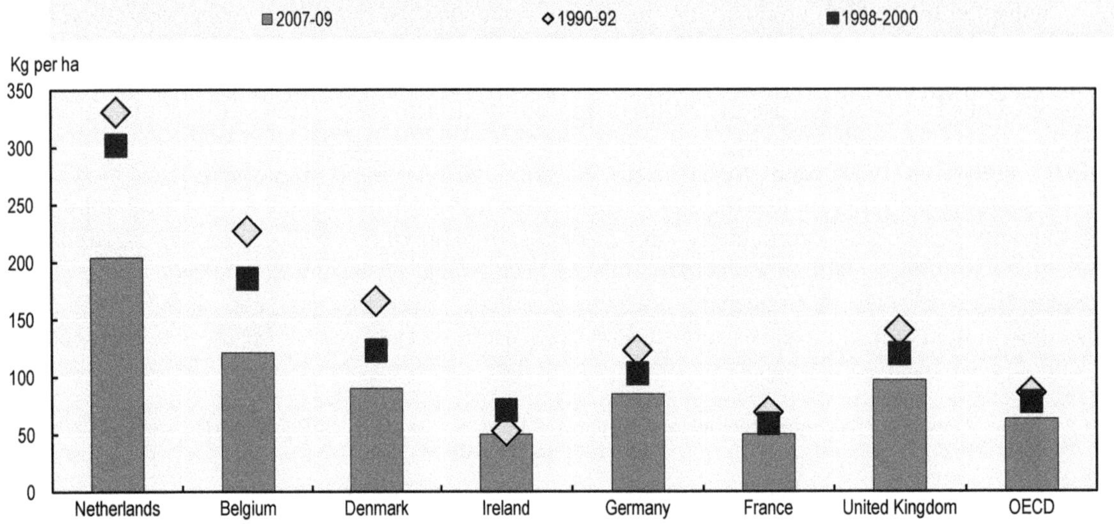

Source: OECD Agri-environmental Indicators database (AEI), 2013, http://data.oecd.org/searchresults/?q=OECD+Agri-environmental+Indicators+database+ .

StatLink http://dx.doi.org/10.1787/888933250824

Different policy responses have been put in place to reduce nutrient pollution from agriculture since the mid-1980s, ranging from a limitation of animal numbers to constraints on manure application (Chapter 6). Partly through these policy measures, a relatively strong decrease in nutrient surpluses has been achieved, as shown in Figure 2.10. Between 1990 and 2012, the Dutch GNB decreased by an average 8.5 kg N/ha per year. The annual rate of decrease has been particularly high under the Mineral Accounting System policy phase between 1998 and 2005. Substantial reductions in the phosphorous balance per hectare of agricultural land also took place over the same period. Such a decrease in nutrient surplus is essentially attributable to a strong reduction in nutrient inputs, combined with a relatively smaller decrease of nutrient outputs. Thus, *nitrogen efficiency*, i.e. the ratio of nitrogen outputs over inputs, equalled 53% in 2012, compared to a value of about 45% in the 1990s (Figure 2.11). Overall, this is in line with the trend observed for TFP overall, which shows that the Dutch agriculture has moved from a period of "high output growth with contained input growth" (1961-90) to a period of "stable output with decreasing input" (since 1990).

In spite of the important progress in reducing environmental pressures due to nutrients, the policy challenge regarding agricultural nutrients in the Netherlands is to continue in the future to improve nutrient efficiency in order to reduce the associated environmental pressures, while at the same time ensuring the competitiveness and productivity growth of the sector (Oenema et al., 2014). This challenge concerns both nitrogen and phosphorous, although in recent years, the phosphorus surplus continued to decline at a rapid pace, unlike the nitrogen surplus, which no longer decreases significantly. It is noticeable that for the recent years – since 2005 – TFP of Dutch agriculture has continued to grow at a steady pace, and even accelerated, while nitrogen efficiency has been stagnating (Figure 2.3 and Figure 2.5). This suggests a potential disconnection between productivity growth and gains in environmental productivity, at least for nutrients in recent years, which raises the issue of the capacity of Dutch agriculture to ensure sustainable productivity growth in the future.

Figure 2.10. Nitrogen inputs, outputs and Gross Nitrogen balance per ha of agricultural land area in the Netherlands, 1998-2005

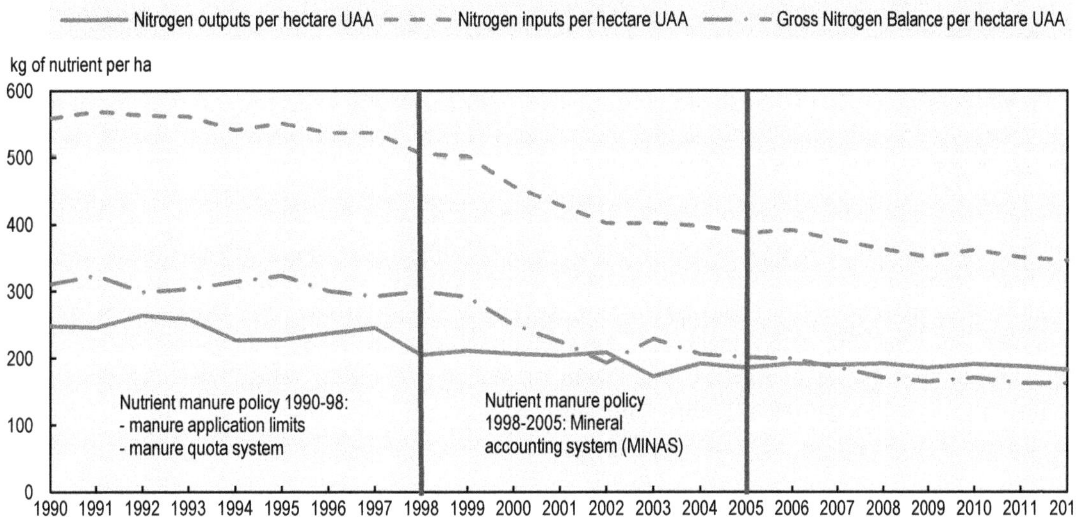

Source: Eurostat (extracted in February 2015).

StatLink http://dx.doi.org/10.1787/888933250835

Figure 2.11. Nitrogen use efficiency in agriculture, Netherlands, 1990-2012

Outputs divided by inputs

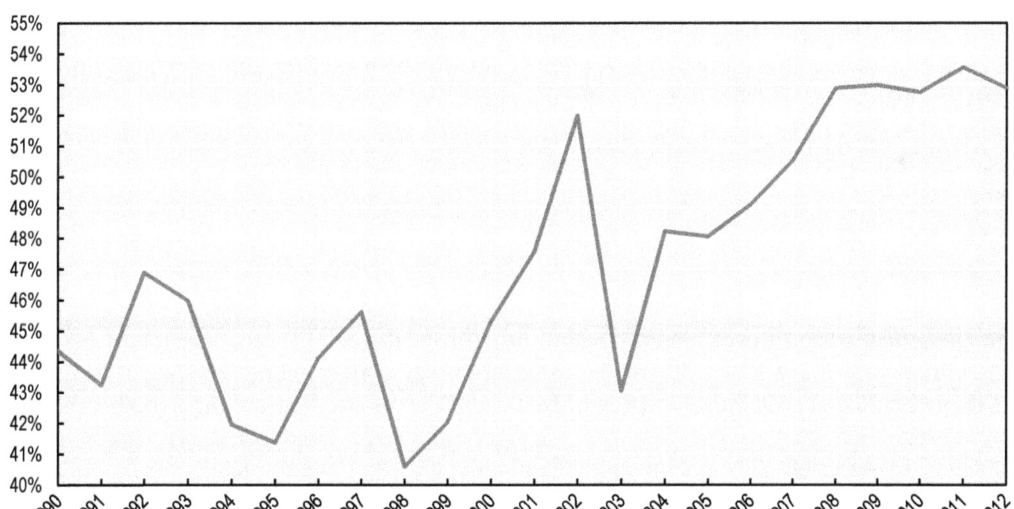

Source: Calculated from the Eurostat Agri-environmental Indicators database (2015), http://ec.europa.eu/eurostat/web/agri-environmental-indicators.

StatLink http://dx.doi.org/10.1787/888933250848

At the same time, indicators of the state of water quality indicate that a significant share of watersheds still do not satisfy the criteria for environmental compliance in terms of Environmental Quality Standards (EQS) for nutrients (Rozemeijer et al., 2014), although the trend is still declining.[5] Excessive nutrient concentrations are then associated with low quality of water and eutrophication of water bodies, although there is a great deal of variations across locations and classes of water bodies (Van Puijenbroek et al., 2014). Such slowing trends in environmental efficiency improvement and the need to meet water quality standards suggest a potential need for innovative technical, technological and policy responses.

In link with the reduction of nitrogen balance, the Netherlands has achieved a great reduction in ammonia emissions. Ammonia emissions in 2012 were 6% lower than the target ceiling set out in the Gothenburg protocol for year 2010,[6] and the reduction continues. Approximately 85% of the national ammonia emissions come from the agricultural sector. Besides through the reduction in livestock numbers enforced through policy measures, ammonia emissions declined in the early 1990s primarily through the compulsory low-emission application of manure. In recent years, the emergence of low emission stalls has had a particularly important role.

Pesticides

The use of chemical pesticides in Dutch agriculture and horticulture roughly halved between the mid-1980s and the turn of the century. The number of authorised products was reduced over the course of the 1990s. Over the past ten years, however, a gradual increase has taken place once again, from 9.6 million kg in 2010 to 11.36 million kg of active substance in 2012, partly due to wet conditions. The ambition of national policy in the Netherlands in this area is "to reduce the number of violations of water quality standards by 90%", under the Second Memorandum on Sustainable Crop Protection (Berkhout et al., 2014)

Some studies suggest there may be significant room to improve the technical efficiency of pesticide use on Dutch arable farms, with potential gains for both farmers and the environment (Skevas et al., 2012; Skevas et *al.*, 2014). However, these empirical evidences also show improvements in terms of technical change for the period 2004-07. These countervailing trends might be explained by some kind of adjustment process: as farmers adopt new techniques and practices, there is a learning-by-doing period that may lead to reduced inefficiencies in the first years following adoption, after which they vanish.

Energy

The Netherlands has an exceptional position in terms of the energy consumption of its agriculture and horticulture, with a share of 6.3% of the national energy consumption. This stems from the large scale of its greenhouse horticulture. Improvements in efficiency are important for both the sector and the government. The goal of the *Kas als Energiebron* (KaE, Greenhouse as source of energy) programme is to make all new greenhouses climate-neutral and economically profitable from 2020 onwards (Box 2.2). Despite the increase in the use of sustainable energy, the objective of 4% by 2010 set out in the Agro Covenant was not achieved.

Agriculture and horticulture account for approximately 12% of the total greenhouse gas emissions in the Netherlands. Since the mid-1990s, emissions from primary agriculture and horticulture have been steadily declining; however, they have risen again slightly since 2007. This is a consequence of the increased CO_2 emissions in greenhouse horticulture, both from the cultivation itself and from electricity generation. The Netherlands has set itself the goal of reducing overall emissions by 20% by 2020 compared with 1990. This is in line with the EU reduction objective. The objective for primary agriculture and horticulture is a reduction of emissions by 16% in 2020 compared with 1990. In 2010, emission levels were 12.5% less than the 1990 level.

Box 2.2. Innovation and Action Programme "Greenhouse as Source of Energy"

The greenhouse sector in the Netherlands is a major user of gas and electricity, with energy representing 20% to 25% of production costs. Improving energy efficiency in greenhouses is therefore very relevant for supply security due to the price volatility of fossil fuels, the need to produce less CO2 and other environmental issues.

"Greenhouse as Source of Energy" is a public-private partnership launched in 2005, which includes the Dutch Horticultural Growers Association (LTO Glaskracht Nederland) and the Dutch Ministry of Economic Affairs. It aims to make new greenhouses climate neutral and near-independent of fossil fuel, by 2020 and a source of sustainable energy supply by 2050, by developing solutions that are economical affordable and ensure profitability. The main target is to reduce CO_2 emissions by more than half compared with 1990, and reduce energy use per unit of product by 2% every year. As a result 20% of the energy consumed in Dutch greenhouses will come from sustainable energy sources.

The project governance is headed by the Board of LTO Glaskracht Nederland (LTO GK-NL) and the ministry. The management team is the chairman of LTO GK-NL and a representative of the Ministry of Economic affairs. The project team consists of members from the producer organisation and the ministry. In addition, the advisory group and the growers group provide feedback on research proposals. Common principles guide the governance through a shared vision and a Frontrunners strategy, which consists of a smaller group that can share results from achievements. Research management is demand driven with equally shared financing between public and private sector. Control is shared between the growers' organisation and the ministry. It was noted that leadership from the producer organisation was found to be instrumental.

A range of policy instruments exists including support to R&D transfer and communication, laws and regulations, loan guarantees (for geothermal projects), subsidy on investment, tax reduction for energy saving investments and an exploitation subsidy for renewal energy.

Results of this work as of 2014 have shown that energy efficiency has doubled since 1990, more geothermal energy sites are up and running, and more closed or semi-closed greenhouses are in use among growers. Diffuse glass has a proven effect on CO_2 reduction, Starting new innovations include new greenhouses that are climate neutral, new dehumidifying options, and hybrid LED options and photosynthesis sensors.

Combined heat and power systems (CHPs) can produce electricity as well as heat from a cubic metre of natural gas. Moreover, CO_2 that is emitted in industry can be used by growers in their greenhouses to stimulate growth. A pipe has already been laid between the Botlek area to greenhouses in the Westland and other areas for this purpose. Furthermore, growers use the CO_2 that is emitted from their own CHPs and heating boilers as much as possible.

Source: Mourits, J. (2014), "Innovation & action program Greenhouse as Source of Energy", presentation at a meeting of the OECD Network for Food Chain Analysis on PPPs for agricultural innovation, October 2014, Paris, available at: http://www.oecd.org/site/agrfcn/6th-oecd-food-chain-analysis-network-meeting-october-2014.htm, and http://www.energiek2020.nu/fileadmin/user_upload/energiek2020/docs/Algemeen/Leaflet_Engels.pdf.

Biodiversity

In the Netherlands, ecosystems are characterised by a high level of anthropisation. Agricultural land accounts for 43% of total land area, which is significantly lower than the EU15 average (60%). Intensive agricultural systems are sources of harmful pressures to ecosystems, due to soil acidification, eutrophication and loss of natural habitats (OECD, 2008; OECD, 2013). The share of agricultural land in total land area has been fairly stable for the last two decades, with a relative decrease of the share of permanent pastures in favour of croplands. The share of land organically-farmed has increased continuously over the last two decades, from 0.4% of all agricultural land in 1990 to 1.6% in 2000 and 2.5% in 2012. Impacts of such change in agricultural land cover on biodiversity and ecosystem services are complex to characterize and there is a lack of data in this area.

Of particular concern is the huge reduction of farmland birds, which as "indicator species" provide a proxy of the health of ecosystems. Recent estimates show a huge decrease of this index for the Netherlands (Figure 2.12).

Figure 2.12. Trends in populations of farmland birds in a set of OECD countries, 1990-2010

Annual average percentage change

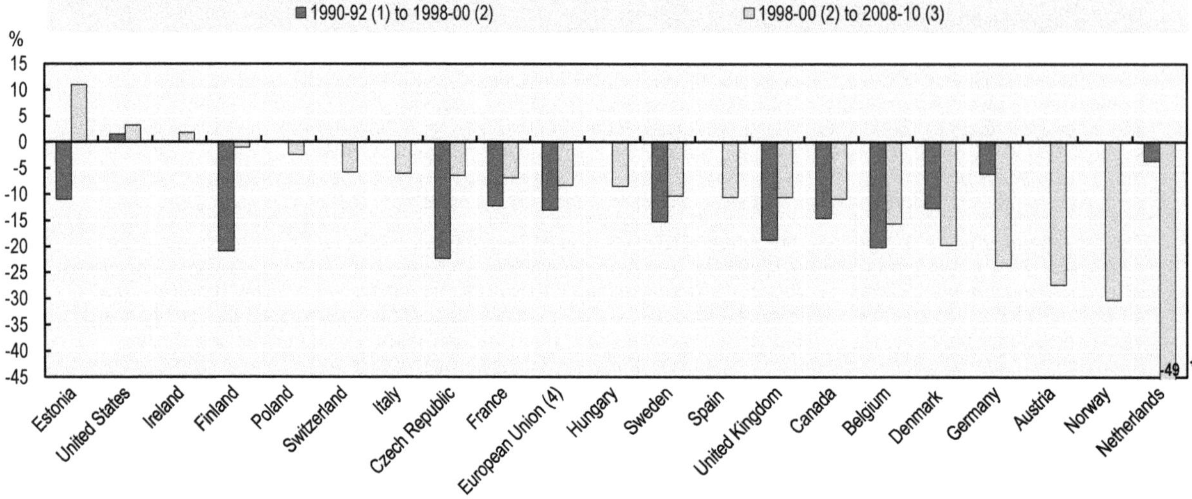

Aggregated index of population trend estimates of a selected group of breeding bird species that are dependent on agricultural land for nesting or feeding. For Canada and the United States these are only grassland breeding birds.

Countries are ranked in descending order according to average annual percentage change 1998-2000 to 2008-2010.

1. Data for the 1990-92 average equals 1991-93 average for Germany.

2. Data for the 1998-2000 average equals 1999-2001 average for Hungary and Switzerland; and 2000-02 average for Italy and Poland.

3. Data for the 2008-10 average equals 2004-06 average for Estonia; 2005-07 average for the United States, 2007-2009 average for Hungary and 2006-08 average for Belgium, Denmark, Finland, France, Germany, Ireland, Spain, Sweden, Switzerland and United Kingdom.

4. The EU aggregate figure is an estimate based on the following 17 Member States: Austria, Belgium, the Czech Republic, Denmark, Estonia, Finland, France, Germany, Hungary, Ireland, Italy, Latvia, the Netherlands, Poland, Spain, Sweden, and United Kingdom.

Source: OECD Compendium of Agri-environmental Indicators, 2013, http://data.oecd.org/searchresults/?q=OECD+Compendium+of+Agri-environmental+Indicators.

StatLink http://dx.doi.org/10.1787/888933250859

There is a lack of knowledge about the exact reasons for the decrease in farmland birds in the Netherlands. Acidification of habitats associated with nutrients surplus; dessication; and the use of pesticides could play a role in this decreasing trend (OECD, 2008). Knowledge gap in this area is a key limiting factor to undertake appropriate policy responses to ensure that agricultural growth is compatible with biodiversity conservation and habitats. This is certainly an area that still requires investment in Research and Development. There is evidence, however, that in several areas the learning process, and increasing co-operation between farmers, ornithologists and nature associations, have contributed to halting the decline of certain species, such as meadow birds in Ronde-Hoep (Noord-Holland) and Arkemheen-Eemland.

Cost and benefits associated with environmental externalities in the Netherlands

Agriculture in the Netherlands is a source of both negative and positive externalities. Assessing the relative importance of these externalities in the Dutch economy requires some estimates of their monetary costs. To date, only a few estimates are available, notably on the cost of water pollution due to nitrogen and phosphorous surplus (Moxey, 2012). For instance, Howarth et *al.* estimated the total cost of nutrient pollution to be in the range between 400 and EUR 750 million. A recent study by Polman et al. (2014) aims at providing a first assessment of the aggregate monetary costs associated with negative externality from agriculture, which was estimated to one-third of the value-added of agriculture at the country scale. Although such monetary valuation exercises are subject to numerous caveats and should be interpreted with caution, the few studies in this area suggest that the environmental costs of agriculture do indeed represent a significant share of the value-added of the sector, providing policy rationale for policy responses and innovation strategies targeted to the sustainable dimension of agriculture.

Climate change and sustainable productivity

Climate change can affect agricultural productivity in two ways: first, mitigation efforts can be costly to implement for farming systems, thus affecting production scale and farming practices; second, climate change will have direct impacts on agricultural production; and indirect impacts through adaptation to changing climate conditions.

To date, knowledge of the projected impacts of climate change on the Dutch agriculture remains uncertain (Stoorvogel, 2009). Several studies suggest that climate change may in fact benefit agricultural production, due to increasing productivity associated with higher CO_2 concentrations and temperature. However, more complex effects of climate change, notably on seasonal rainfall patterns and pests and diseases complicate projection exercises, since there is a lot of uncertainty in these areas. In general, impacts of climate change on livestock are less understood than impacts on crops; in particular because they are more indirect – through changes in pasture productivity, developments of pests and diseases, etc. (OECD, 2014). Higher risks of floods and water shortages could be expected, with impacts on agriculture but also beyond, at the wide-economy-level.

Challenges for the future

The sector operates in a wealthy and densely-populated country, which means that the costs of land and labour are fairly high. Continuous innovation has been the main driver of the competitiveness of Dutch agri-food sector on foreign markets. Facing natural resource and regulatory constraints, and high input costs, in particular land and labour, the sector has achieved high levels of productivity for all factors. Prior to 1990, productivity growth was achieved by both intensification and growth of factor use leading to output increasing faster than input use. From the 1990s TFP growth was mainly driven by higher efficiency in the use of inputs and the reduction of input use did not affect production levels. As a result sustainability performance has improved but significant efforts are still required in this area.

The sector continues to face resource and cost constraints, but with changes in intensity. For example, some costs have increased or become more variable due to the economic crisis, instability in energy markets, stricter regulations (on animal welfare, fertiliser and pesticide use and animal health products). As the size of operations increases, the consequences of livestock diseases become larger.

The sector is exposed to the increasing demands of society with respect to production methods and the resulting products. The negative environmental externalities and new public concerns like those about animal welfare and the quality of nature and the rural area, as well as competition from foreign suppliers, are all demanding agriculture to change in new directions. The sector is facing the threefold PPP challenge: to ensure continued income (profit) for the efficient producers; to offer quality products at reasonable prices for the people; to protect the natural capital of the planet. To meet this multiple challenge, a transition towards sustainable agriculture is called for. Producers (farmers and horticultural growers), the agri-food trade and industry, the retail sector and governments all need to deal with this challenge.

A major challenge regarding the impacts of climate change on Dutch agriculture, and the associated needs for adaptation responses, is the uncertainty about the exact nature and magnitude of these impacts. This is a general issue which is not specific to Netherlands: the assessment of climate change impacts on agriculture is plagued by a cascade of uncertainties, which tend to multiply when one moves from direct to indirect effects; and from low to high geographical scale (OECD, 2013).

Agricultural policy changes offer the scope for a better adaptation of the sector to its changing environment. EU markets are less distorted with the reduction of public intervention, the removal of dairy quotas (and soon of sugar quotas), the replacement of most commodity-linked payments by payments, which have minimal impact on production (production is not required), and will be gradually more equally distributed. The Common Agricultural Policy also offers more opportunity to target payments to specific objectives such as investment in innovative technologies, adoption of sustainable practices or risk management tools. The commonly agreed and applied Sanitary and Phytosanitary regulations also help improve the functioning of the EU markets.

Foreign markets also offer opportunities with higher demand due to population and income growth, and the segmentation of markets for products with specific health or environmental attributes.

The agricultural innovation system will have to supply solutions that help save resources and energy, while improving the productivity and quality of agri-food products and at the same time minimising negative externalities on the environment. This requires a long-term strategy and increased co-operation at national and international levels. The government has traditionally played an important role in driving and funding of agri-food R&D and education, while being responsive to industry demand and developing partnerships. This has permitted the development of solutions addressing both long-term and short-term challenges. Although declining, government funding still accounts for the large majority of investment in agricultural R&D, but the government no longer drives the research and innovation agenda. A challenge for the future of innovation policy is to find the balance between support to business-led innovation with marketable results and support to R&D to deliver solutions for public good and longer-term issues.

Another issue more specific to the Netherlands, with its high population density, high environmental constraints, and well-developed food-complex, is the relationship between domestic primary agriculture and agri-food R&D. The more specific question raised by different bodies is whether the Netherlands would better specialise as an experimenter and developer country for agriculture rather than a commodity production country, whether Dutch R&D capacity would then remain in the country or move abroad, and whether innovation without a domestic production base can be successful.

Summary

- The Dutch food, agricultural and horticultural sector is diverse, innovative, and well-integrated in global agro-food chains. It generates high-value added, and is competitive on world markets.
- The high levels of land and labour productivity in primary agriculture result from the widespread and continuous adoption of innovation that has increased input use efficiency and the continuous process of rationalisation, consolidation, mechanisation and specialisation. The reduction of input use in the last decade has improved the environmental performance of agriculture, but production systems continue to impose pressure on natural resources.
- The sector faces relatively high land and labour costs, as well as increasing environmental and animal welfare constraints reflecting society's demands. It also faces higher uncertainties related to policy changes, general economic and agri-food market developments, and climate change. At the same time competition on world markets becomes stronger.
- In this context, fast innovation is required to remain competitive. The agricultural innovation system will have to supply solutions that help save resources and energy, while improving the productivity and quality of agri-food products.

Notes

1. The OECD regional typology classifies Territorial Level (TL)3 regions as predominantly urban, intermediate or predominantly rural. A region is classified as urban if the share of the population living in rural local units is below 15%. It is classified as intermediate: 1) if the share of population living in rural local units is between 15% and 50%, and if it does not contain an urban centre of more than 500 000 inhabitants representing at least 25% of the regional population; or 2) if the share of population living in rural local units is over 50%, and if it contains an urban centre of more than 200 000 inhabitants representing at least 25% of the regional population. See definition of typology at: www.oecd.org/gov/regional-policy/OECD_regional_typology_Nov2012.pdf. None of the Dutch regions at TL 3 are defined as predominantly rural, seven are predominantly urban and five are intermediate. See OECD regional database at: www.oecd.org/gov/regional-policy/regionalstatisticsandindicators.htm.

2. The sub-complexes include the input supply, primary production, processing and distribution connected in the value chain. The shares of each sub-complex are calculated using Input-Output matrices.

3. OECD WTO Trade in Value-Added Database (TiVA), 2013. www.oecd.org/industry/ind/measuringtradeinvalue-addedanoecd-wtojointinitiative.htm.

4. These figures come from the Eurostat database on agri-environmental indicators (http://ec.europa.eu/eurostat/web/agri-environmental-indicators/overview). Due to recent methodological changes in Dutch statistics, Eurostat data on agricultural nitrogen and phosphorous balances may differ from Statistics Netherlands. Notably, according to Statistics Netherlands, the nitrogen surplus (N kg per hectare) has been estimated at 230 kg/ha in 1998 (76 kg per ha for the phosphate surplus); and at 113 kg/ha for 2012 (14 kg per ha for the phosphate surplus) (Berkhout et al., 2014). However, these methodological variations do not question the declining trend of mineral surpluses since 1990. To ensure comparability across countries, the figures on nutrients presented in this report come from the Eurostat and OECD databases, which are consistent. For more information on Statistics Netherlands data, see Environmental data compendium at: www.compendiumvoordeleefomgeving.nl/dossiers/nl0225-Environment.html?i=41-205.

5. Changes in environmental pressures can have lagged impacts on the state of the environment. Notably, it can take several years for nutrient policy measures to translate into improved water quality for surface and ground water. Hence links between pressures and states of the environment should take these time-lags into consideration.

6. The 1999 Gothenburg Protocol to Abate Acidification, Eutrophication and Ground-level Ozone "sets emission ceilings for 2010 for four pollutants: sulphur, NOx, VOCs and ammonia. These ceilings were negotiated on the basis of scientific assessments of pollution effects and abatement options. (...) The Protocol was amended in 2012 to include national emission reduction commitments to be achieved in 2020 and beyond". (Source: www.unece.org).

References

Berkhout, P., H.J. Silvis, and I. Terluin (eds.) (2014), *Agricultural Economic Report 2014 of the Netherlands: Summary*, LEI Report 2014-014, LEI Wageningen UR, The Hague, August. Full report in Dutch available at: http://edepot.wur.nl/306953.

Bijman, J., G. van der Sangen, K.J. Poppe, and B. Doorneweert (2012), *Support for Farmers' Co-operatives; Country Report The Netherlands*, Wageningen: Wageningen UR. http://edepot.wur.nl/244818 .

Eurostat (2015), Eurostat Agri-environmental Indicators database, http://ec.europa.eu/eurostat/web/agri-environmental-indicators.

Eurostat (2014), Newsrelease 194/2014, 15 December: http://ec.europa.eu/eurostat/documents/2995521/6330108/5-15122014-BP-EN.pdf/52f8f552-9faf-406f-80ef-3ca01e45e2e1.

Jongeneel, R., Polman, N., & Ham, C. Van Der. (2014), "Costs and benefits associated with the externalities generated by Dutch agriculture", Working Paper, pp. 1–14.

Kimura and Sauer (2015), "Dynamics of Dairy Farm Productivity Growth: Cross-country Comparison", *OECD Papers on Food, Agriculture and Fisheries*, OECD Publishing, Paris (forthcoming).

Miroudot, S. (2014), "Trade in value-added in the Agriculture and Food Industries", *EuroChoices*, Vol. 13, Number 2, pp. 28-29.

Mourits, J. (2014), "Innovation & action program Greenhouse as Source of Energy", presentation at a meeting of the OECD Network for Food Chain Analysis on PPPs for agricultural innovation, October 2014, Paris, available at: http://www.oecd.org/site/agrfcn/6th-oecd-food-chain-analysis-network-meeting-october-2014.htm.

Moxey, A. (2012), "Agriculture and Water Quality: Monetary Costs and Benefits across OECD Countries", OECD, Paris, http://www.oecd.org/tad/sustainable-agriculture/49841343.pdf .

OECD (2014a), *OECD Economic Surveys: Netherlands 2014*, OECD Publishing, Paris. DOI: http://dx.doi.org/10.1787/eco_surveys-nld-2014-en.

OECD (2014b), *OECD Territorial Reviews: Netherlands 2014*, OECD Publishing, Paris. DOI: http://dx.doi.org/10.1787/9789264209527-en.

OECD (2014c), *Climate Change, Water and Agriculture: Towards Resilient Systems*, OECD Studies on Water, OECD Publishing, Paris. DOI: http://dx.doi.org/10.1787/9789264209138-en.

OECD (2013), *OECD Compendium of Agri-environmental Indicators*, OECD Publishing, Paris. DOI: http://dx.doi.org/10.1787/9789264181151-en.

OECD (2011), *Evaluation of Agricultural Policy Reforms in the European Union*, OECD Publishing, Paris. DOI: http://dx.doi.org/10.1787/9789264112124-en.

Oenema, O., C. De Klein, and M. Alfaro (2014), "Intensification of grassland and forage use : driving forces and constraints", *Crop and Pasture Science*, Vol. 65, pp. 524–537. DOI: http://dx.doi.org/10.1071/CP14001.

Rozemeijer, J. C., J. Klein, H.P. Broers, T.P. van Tol-Leenders, and B. van der Grift (2014), "Water quality status and trends in agriculture-dominated headwaters; a national monitoring network for assessing the effectiveness of national and European manure legislation in The Netherlands", *Environmental Monitoring and Assessment*, Vol. 186, No. 12, pp. 8981–95. DOI:10.1007/s10661-014-4059-0.

Silvis, H.J., C.M. van der Heide, and P.W. Blokland (2013), "De Nederlandse landbouw op het OECD-milieuscorebord", LEI-nota 13-081. http://edepot.wur.nl/276449.

Skevas, T., A.O. Lansink, and S.E. Stefanou (2012), "Measuring technical efficiency in the presence of pesticide spillovers and production uncertainty: The case of Dutch arable farms", *European Journal of Operational Research*, Vol. 223, No. 2, pp. 550–559. DOI:10.1016/j.ejor.2012.06.034.

Skevas, T., and A.O. Lansink (2014), "Reducing Pesticide Use and Pesticide Impact by Productivity Growth: the Case of Dutch Arable Farming", *Journal of Agricultural Economics*, Vol. 65, No. 1, pp. 191–211. DOI:10.1111/1477-9552.12037.

Snijders, H., and D. Jacobs (2013), *Clusters en niches; de specialisatie van de Nederlandse economie*, WRR (Scientific council for government policy), webpublicatie No. 76. http://www.wrr.nl/fileadmin/nl/publicaties/PDF-webpublicaties/76_Clusters_en_niches._De_specialisatie_van_de_Nederlandse_economie.pdf.

Stoorvogel, J. J. (2009), *Adapting Dutch agriculture to climate change*, Knowledge for Climate Report, KfC Report Number KfC 016/09, Wageningen UR.

USDA (2014), Economic Research Service Agricultural Productivity database, consulting in November 2014. Available at: www.ers.usda.gov/data-products/international-agricultural-productivity/documentation-and-methods.aspx#excel.

Van Puijenbroek, P. J. T. M., P. Cleij, and H. Visser (2014) "Aggregated indices for trends in eutrophication of different types of fresh water in the Netherlands", *Ecological Indicators*, Vol. 36, pp. 456–462. DOI:10.1016/j.ecolind.2013.08.022.

Wamelink, G.W.W, B. de Knegt, R. Pouwels, C. Schuiling, R.M.A. Wegman, A.M. Schmidt, H.F. van Dobben, M.E. Sanders (2013), "Considerable environmental bottlenecks for species listed in the Habitats and Birds Directives in the Netherlands," *Biological Conservation*, Vol. 165, pp. 43-53.

Annex 2.A1

Background table

Table 2.A1.1. Share of the agricultural complex in the economy, 1995, 2005, 2011

Sub-complexes	Value added			Employment			Energy use[3]		CO_2 emission[3]	
	1995	2005	2011	1995	2005	2011	2005	2011	2005	2011
Arable farming	18.0	17.1	20.7	19.3	16.4	16.9	9.3	11.3	11.9	10.3
Open field horticulture	8.6	8.0	7.0	9.2	9.7	9.9	2.9	3.4	3.9	5.8
Glasshouse horticulture	18.4	22.0	22.2	14.7	17.5	19.0	52.9	51.7	26.1	28.4
Grassland based livestock farming	33.6	30.2	29.9	34.9	35.3	33.7	17.1	17.3	35.9	34.7
Intensive livestock farming	18.2	21.3	19.3	18.0	19.9	19.6	13.9	12.8	20.1	19.2
Fishery	3.3	1.3	0.9	3.8	1.0	0.9	4.0	3.5	2.3	1.6
Agricultural complex	100	100	100	100	100	100	100	100	100	100
% domestic agricultural complex in national total[1]	*7.7*	*5.1*	*4.9*	*7.9*	*6.1*	*5.5*	*10.0*	*9.7*	*19.0*	*19.5*
% total agricultural complex in national total[2]	*12.3*	*9.8*	*9.8*	*12.1*	*10.3*	*10.0*	*14.3*	*14.4*	*23.1*	*23.8*

1. Based on domestic agricultural resources (excluding cocoa, drinks, tobacco).

2. Based on domestic and imported agricultural resources.

3. 1995 figures for energy use and CO_2 emission not available.

Source: Berkhout, P., H.J. Silvis, and I. Terluin (eds.) (2014), *Agricultural Economic Report 2014 of the Netherlands: Summary*, LEI Report 2014-014, LEI Wageningen UR, The Hague, August. Full report in Dutch available at: http://edepot.wur.nl/306953.

StatLink http://dx.doi.org/10.1787/888933251342

Chapter 3

Economic stability and quality of institutions in the Netherlands

This chapter gives an overview of the performance of the overall economy, macroeconomic developments and challenges, and the governance and institutions.

The statistical data for Israel are supplied by and under the responsibility of the relevant Israeli authorities. The use of such data by the OECD is without prejudice to the status of the Golan Heights, East Jerusalem and Israeli settlements in the West Bank under the terms of international law.

Macro-economic policy environment

At the broadest level, stable and sound macroeconomic policies play an important role in setting a favourable environment for investment by farms or agri-food firms seeking to introduce new products, to adopt new production methods, or to undertake organisational changes that can lead to higher productivity growth and more sustainable use of natural resources (OECD, 2014a).

Performance of the overall economy

The Netherlands is a country with high standards of living in almost all domains. According to the OECD better life index, the country is performing particularly well with respect to work and life balance and life satisfaction; and in line with the OECD average in terms of safety, environment and civic engagement.[1] The global crisis has, however, severely affected this export-oriented economy, which is still recovering.

In the top of world competitiveness

The World Economic Forum (WEF) Global Competitiveness Indicators for 2013-14 (WEF, 2013) rank the Netherlands at the 8th place out of 148 (Figure 3.1).[2] Overall, the Netherlands is regarded as a well-developed and highly-productive knowledge economy, based on some pronounced strengths:

- The **innovation system** is performant: The higher education sector produces world-class science and the relative number of patents (Patent Co-operation Treaty, PCT) filed by universities and public labs is above the OECD median (OECD, 2014a). Dutch businesses are highly sophisticated (4th) and innovative (10th), and the country is rapidly and aggressively harnessing new technologies for productivity improvements (8th). The Netherlands has a well-performing knowledge economy relative to investments, as the economy remains innovative and competitive despite the decline in R&D investment (Chapter 7).
- The **education system** is considered excellent. It ranks 4th for health and primary education and 6th for its higher education and training according to WEF Global Competitiveness Indicators (Chapter 5).
- The **labour force** is well-educated and skilled, including in problem solving and science; tertiary education yields better employment prospects in the Netherlands than in OECD countries on average, but the employment prospects of secondary education is also higher than the OECD average. Labour markets are regarded as relatively efficient, compared to the OECD average, but less liberal than in some other countries (20th rank) (Chapter 5).
- **Markets** are considered efficient and highly supportive of business activity. The Netherlands ranks 8th for goods markets in particular.
- **Institutions** are considered as highly reliable (8th rank) (Chapter 3).
- Last but not least, the quality of its **infrastructure** is among the best in the world, reflecting excellent facilities for maritime, air, and railroad transport, which are ranked 1st, 4th, and 11th, respectively (Chapter 5).

The **macroeconomic environment** does not rank as high as other competitiveness indicators as the economy is still recovering from a post-crisis recession and suffers from weak productivity growth (OECD, 2014b). The environment is, however, stronger than that of a number of other advanced economies (see next section). Financial market development is also an area with relatively lower ranking (30th), even though it compares well with the OECD average (Figure 3.1.C). Respondents to the WEF survey indicate that access to financing, restrictive labour regulations and inefficient government bureaucracy are the three most problematic factors for doing business (Figure 3.1.C).

Figure 3.1. Indicators of Dutch global competitiveness and problematic factors for doing business

A. Global Competitiveness Index, score by component, 2013-14

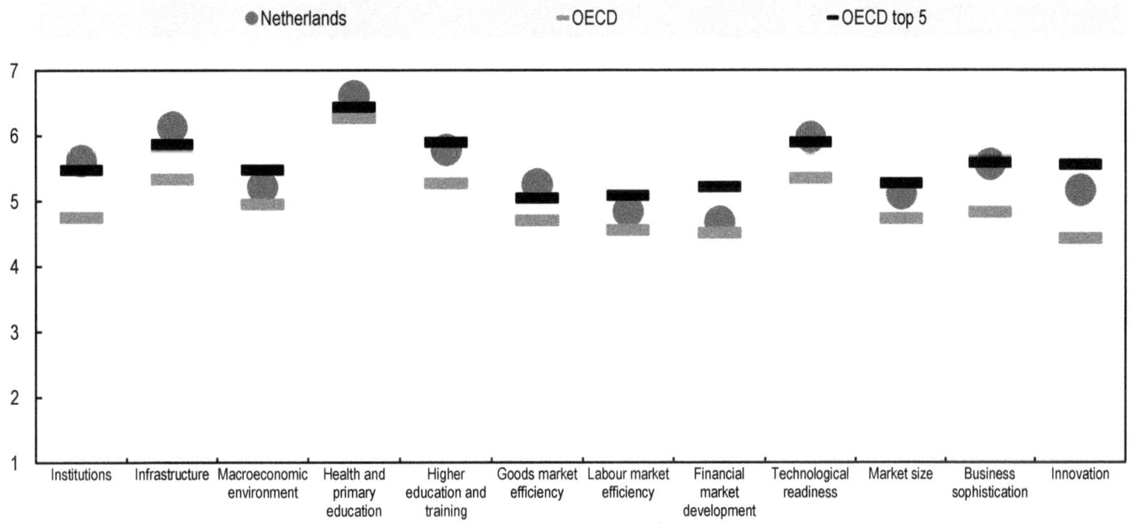

B. Global Competitiveness Index, rank by component, 2013-14

		Efficiency enhancers (50%)	**11**
GCI 2013-14 (out of 148)	8	Higher education and training	6
GCI 2012-13 (out of 144)	5	Goods market efficiency	8
GCI 2011-12 (out of 142)	7	Labour market efficiency	21
Basic requirements (20%)	**10**	Financial market development	30
Institutions	8	Technological readiness	8
Infrastructure	7	Market size	21
Macroeconomic environment	45	**Innovation and sophistication (30%)**	**7**
Health and primary education	4	Business sophistication	4
		Innovation	10

C. The most problematic factors for doing business

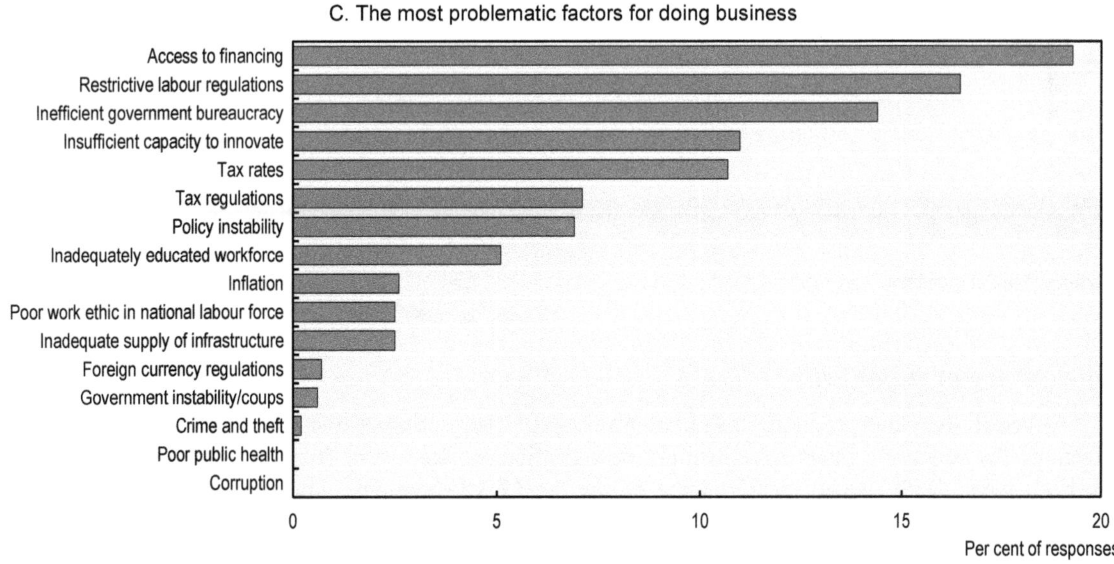

From the list of factors above, respondents were asked to select the five most problematic for doing business in their country and to rank them between 1 (most problematic) and 5.

The bars in the figure show the responses weighted according to their ranking).

Source: World Economic Forum (2013), *The Global Competitiveness Report 2013-2014: Full data Edition*, Geneva 2013. http://reports.weforum.org/the-global-competitiveness-report-2013-2014/#=.

StatLink http://dx.doi.org/10.1787/888933250867

Macroeconomic indicators traditionally strong

Over the long-term, macroeconomic indicators are traditionally strong for the Dutch economy. The country has one of the highest GDP per capita among OECD countries (Figure 3.2). In 2013, per capita GDP (in constant 2005 USD) ranked fifth amongst OECD member countries, following Luxembourg, Norway, the United States and Switzerland. The Dutch economy is one of the most open economies in OECD countries. In 2012 exports accounted for close to 90% of GDP, more than double the OECD average and higher than both Denmark and Germany (OECD, 2014c).

Employment rates (employment to population ratio) in the Netherlands are higher than the OECD average. The rate increased significantly by 11 percentage points over the period 1994-2012, standing at 75.1% in 2012, one of the highest in OECD countries, well above the United States and just above the levels of Germany, Sweden and Denmark. Notwithstanding this fact, average working hours are comparatively low by international comparison. After an important acceleration between 2004 and 2007, the employment rate has remained constant during the subsequent years (2008-12) (OECD, 2014b).

The Dutch economy is, however, facing a number of macro-economic challenges outlined below.

Figure 3.2. Trends in GDP per capita, 1992-2013

USD PPP 2005

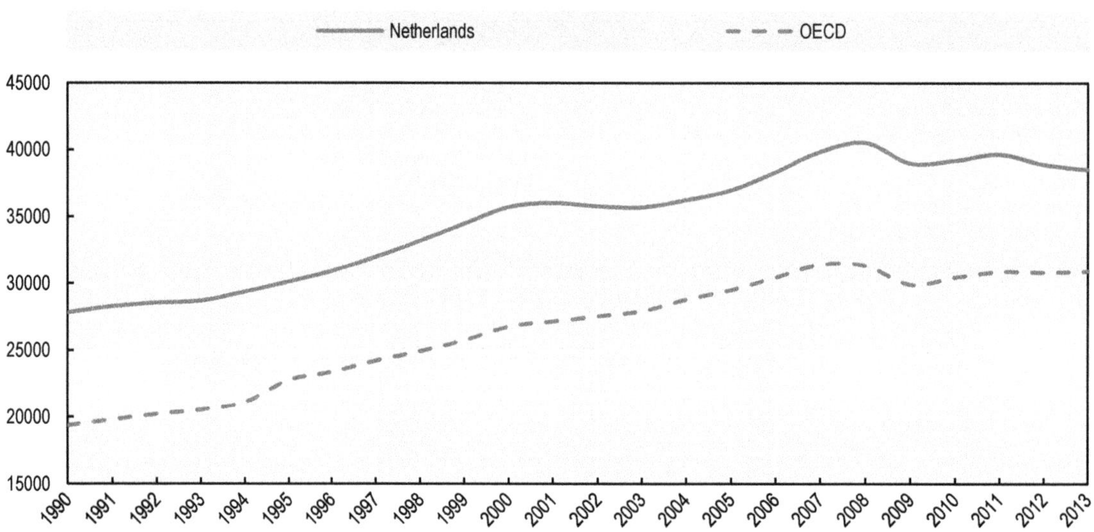

Source: OECD statistics. http://dotstat.oecd.org/.

StatLink http://dx.doi.org/10.1787/888933250878

Macro-economic challenges[3]

Although the Dutch economy has been traditionally very competitive among OECD countries, the effects of the economic crisis have brought new challenges. Recovery from the global financial crisis has been weak, with growth performance below the OECD average. The growth of GDP has been negative almost uninterruptedly since mid-2011 (Figure 3.2). The unemployment rate, although still below the EU average, has experienced a worrying increase since the financial crisis in 2008, with quarterly data for 2013 higher than the previous decade. Ongoing fiscal consolidation and household deleveraging holds back internal demand, and growth is too weak to prevent a further rise of the unemployment rate. Forward-looking indicators suggest that economic weakness is likely to continue.

Squeeze of credit

The crisis of the banking sector, exacerbated by the collapse of the real estate market, has led to a squeeze of credit, mainly affecting small and medium enterprises (SMEs). Banks are still undergoing a process of restructuring of their balance sheets, that is likely to negatively affect the access to credit in the near future. Access to credit for businesses remains an important bottleneck for the Dutch recovery. SMEs are experiencing more difficulty in access to credit than similar firms in other EU economies, particularly worrying is the share of bank loan applications rejected.

Labour productivity decreasing

Labour productivity is higher than the OECD average, by about 30% in 2013, but has experienced an important drop during the 2009 crisis and the 2011 recession. Over the 2008-13 period, it has recorded an annual decrease of 0.2% per year compared to a 0.8% annual increase in the OECD area (OECD, 2014d).

Rising unemployment

Recent data display an important rise in the unemployment rate. There was a positive decreasing trend before the crisis from 6.5% in 2005 to below 4% in 2008, but this trend is reversed in recent years, reaching again the value of 6.5% in 2013 (OECD, 2014c). The rise of unemployment is occurring across all levels of educational attainments. In 2011, the unemployment rate of workers with low education was close to 5.5%, while the unemployment rate of the highly educated workers was 2.75%.

Ageing of the population

Long term challenges are brought by the ageing of the population and shrinking of the labour force. In terms of population structure, in the year 2011 almost 67% of the national population was working age population (15-65), while the share of young people (0-15) was slightly larger than the share of elderly (over 65), accounting for 17.45% and 15.58%, respectively. Although the share of young people is still larger than the share of elderly people, the dynamics of each component of the population shows a worrying trend. The percentage of elderly (65+) compared to the working population (15-65) lies between the OECD and EU average at present, but is expected to surpass the EU average in 2023 (Martinez-Fernandez et al., 2013, quoted in OECD, 2014e).

Government budget under stress

The government budget has been under stress during the past years. In the period 1996-2007, the deficit was under control leading to a decrease in the level of debt. The financial crisis changed this trend with an increase of the debt/GDP ratio and a level of deficit above the 3% limit set by the EU Stability and Growth Pact. In order to comply with the EU Stability and Growth Pact, a restrictive fiscal policy has been implemented since 2011. As a result, the headline deficit was reduced from about 5.5% of GDP in 2009 to nearly 2.3% of GDP in 2013, and is estimated to reach 2.6% in 2014 and 2.3% in 2015 (OECD, 2014c).

Fiscal sustainability has been strengthened with reforms of the pension system, health care and long-term care. For the housing market the property transfer tax has been lowered since 2011 (from 6% to 2%). Since January 2013, new mortgages are eligible to interest tax deductibility only if they are regularly amortised.

Governance and institutions

Good governance systems and high-quality institutions provide economic actors with the assurance that the government is accountable, transparent and predictable. They are a fundamental pre-condition both to encourage public and private investment in the economy and to enable those investments to achieve the intended benefits, both for investors and the host country. Moreover, governance systems play an important role in addressing market failure, influencing the behaviour of firms as well as the efficient functioning of input and output markets (OECD, 2014a).

Political institutions

The Netherlands, part of the Kingdom of the Netherlands, is a parliamentary constitutional monarchy (Box 3.1).

Box 3.1. The Kingdom of the Netherlands

The Kingdom of the Netherlands has four parts: the Netherlands, Aruba, Curaçao and Sint Maarten. Most of the kingdom affairs are administered by the Netherlands (which comprises roughly 98% of the Kingdom's land area and population) on behalf of the entire kingdom. Aruba, Curaçao, and Sint Maarten have their own parliaments. The vast majority of the constituent country of the Netherlands (as well as the kingdom) is located in Europe, with the exception of its three special municipalities (Bonaire, Sint Eustatius and Saba) that are located in the Caribbean. The countries Aruba, Curaçao and Sint Maarten are also in the Caribbean.

The Dutch Constitution of 1848 established a decentralised unitary state consisting of the government, the parliament (the States-General) and the subnational government level. The national parliament has legislative authority in most policy areas discussed in this report, with the exception of land and water use regulations.

The subnational level comprises two tiers of government with general competencies, the provinces (*provincies*) and the municipalities (*gemeenten*), as well as a functional layer comprising the regional water authorities (*waterschappen*) (OECD, 2014e).

- The provinces are one of the oldest institutions in the Netherlands, and their number and size remained constant for centuries. Flevoland, which mainly consists of reclaimed land, was acknowledged as the newest province of the Netherlands in 1986.
- By contrast, the municipal landscape has undergone profound changes over the years following a continuous process of amalgamations. As of January 2014, there are 403 municipalities (compared with 811 in 1980).
- The 23 regional water authorities manage the country's elaborate system of dykes and polders. They are responsible for the operation and management of regional water systems, flood defence, water quality, wastewater transport and treatment as well as musk rat and coypu control. The water boards are an historical feature of the Dutch administrative system, reflecting the country's particular relationship with water. In the last 50 years, important water governance reforms have consolidated the robust legal and policy framework for the water sector.

All three have deliberative assemblies, which are elected by direct universal suffrage. They have an autonomous power of regulation and administration of their own internal affairs. For the provinces and municipalities, these are based on a territorial division and general competencies ("territorial decentralisation") and for the regional water authorities on a functional division and more specialised competencies ("functional decentralisation"). These subnational authorities also have the power to raise taxes.

Abolition of commodity and industrial boards

For many years, the statutory trade organisations (commodity and industrial boards, in Dutch: *productschappen en bedrijfschappen*) formed another type of functional decentralisation. These institutions have all been abolished as of 1 January 2015. This marked the end of a number of special public institutions formed in the 1950s. The primary duty of these boards was to "serve the public interest by promoting the operations of the businesses for which the boards were formed" in a time in which there was a need to assign organisations of employers and employees part of the responsibility for the country's socio-economic policy. This need was particularly great in the post-war period, when it was necessary to concentrate on the recovery and reconstruction of the country's economy. The commodity boards represented enterprises that work with the same product, from raw material to finished product. Industrial boards represented enterprises that fulfil the same function in the economy, for example all farmers (The "*Landbouwschap*" was abolished in the nineties, after which some of its tasks were taken over by the commodity boards).

The foundation of the European Union and the shaping of European agricultural policy, in particular in the form of market regulations for different products in the 1960s, resulted in an expansion of the duties of the boards and the government's allocation of joint administrative duties to the boards. Until recently the boards have played an important role in the priority setting and (collective) funding of practical agricultural research.

The parliament approved the abolition of the boards in 2011. The mandatory levies, in particular, had undermined support for these bodies among their members. Also structural changes in the agricultural supply chains have contributed to the abolishment. The boards' public duties have now been transferred to the Ministry of Economic Affairs. The business community can opt to make arrangements, for its own account, for the performance of other duties (such as the provision of information, promotion of their sector and the representation of their interests), for example by the formation of sector organisations.

Quality of institutions

The quality of public institutions is generally perceived as very good by Dutch citizens. The Netherlands scores at the top of the ranking in the main indicators of governance. It ranks among the OECD 5 top performers for the quality of public institutions, estimated by the Word Economic Forum. It scores particularly highly on property rights and control of corruption, both fundamental for creating a sound business environment (Figure 3.3).

Figure 3.3. Global Competitiveness Index: Quality of public institutions, 2013-14

Scale 1 to 7 (best)

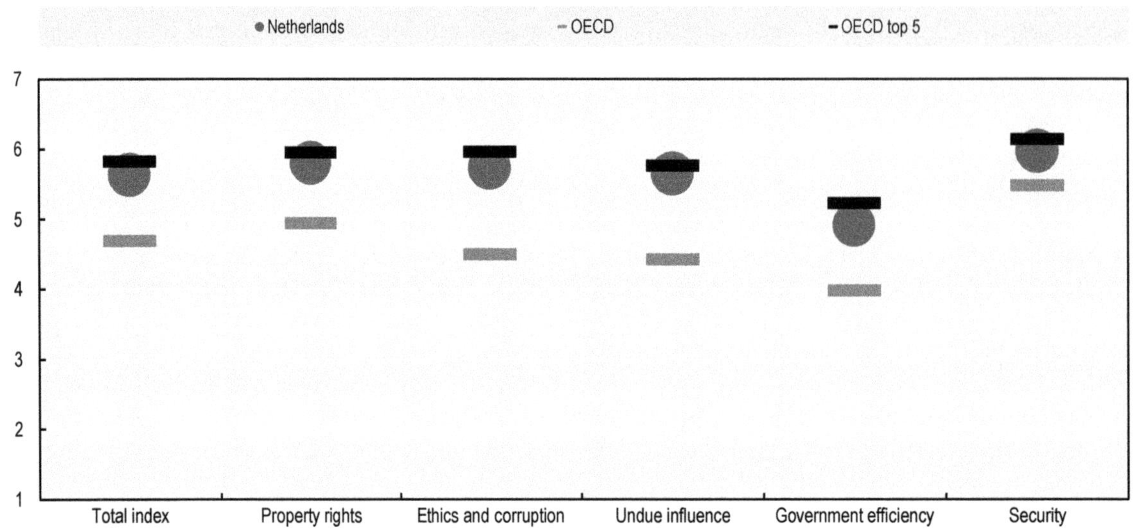

OECD top 5 refers to the average of the scores for the top five performers among OECD countries (Finland, New Zealand, Switzerland, Sweden and Netherlands).

Property rights refer to the average of the indices Property rights and Intellectual property rights. Ethics and corruption refers to the average of the indices: Diversion of public funds, Public trust in politicians and Irregular payments. Undue influence refers to the average of the indices for: Judicial independence and Favouritism in decisions of governmental officials. Government efficiency refers to the average of the indices for Wastefulness of government spending, Burden of government regulation, Efficiency of legal framework in settling disputes, Efficiency of legal framework in challenging regulations and Transparency of government policymaking. Security refers to the average of the indices for: Business costs of terrorism, Business costs of crime and violence, Organised crime and Reliability of police services.

Source: World Economic Forum (2013), *The Global Competitiveness Report 2013-2014*, http://reports.weforum.org/the-global-competitiveness-report-2013-2014/#=.

StatLink http://dx.doi.org/10.1787/888933250886

Summary

- The Netherlands is a high-income, trade-oriented and knowledge-based economy, with sound policy and highly efficient and reliable institutions.
- The world financial and economic crisis has severely hit this open economy, which is gradually recovering from a prolonged recession. The crisis and subsequent government efforts to restore public balance have affected both public and private capacity to invest in innovation.

Notes

1. See OECD web site: http://www.oecd.org/statistics/datalab/bli.htm.
2. After having moved up in the rankings to 5th place, the Netherlands has lost three places and slips to 8th place. The drop mainly reflects weakening financial markets and, in particular, rising concerns regarding the stability of banks.
3. This Section is based on OECD (2014b), *OECD Economic Surveys: Netherlands 2014*, OECD Publishing, Paris. http://dx.doi.org/10.1787/eco_surveys-nld-2014-en.

References

Martinez-Fernandez, C. et al. (2013), "Demographic change in the Netherlands: Strategies for resilient labour markets", *OECD Local Economic and Employment Development (LEED) Working Papers*, No. 2013/13, OECD Publishing, Paris, http://dx.doi.org/10.1787/5k3xnhvzhmxn-en.

OECD (2014a), "Analysing Policies to improve agricultural productivity growth, sustainably: Revised framework". http://www.oecd.org/tad/agricultural-policies/Analysing-policies-improve-agricultural-productivity-growth-sustainably-december-2014.pdf

OECD (2014b), *OECD Economic Surveys: Netherlands 2014*, OECD Publishing, Paris. DOI: http://dx.doi.org/10.1787/eco_surveys-nld-2014-en.

OECD (2014c), *OECD Economic Outlook: Volume 2014/2*, OECD Publishing, Paris. DOI: http://dx.doi.org/10.1787/eco_outlook-v2014-2-en.

OECD (2014d), OECD Science, Technology and Industry Outlook 2014, OECD Publishing, Paris. DOI: http://dx.doi.org/10.1787/sti_outlook-2014-en.

OECD (2014e), *OECD Territorial Reviews: Netherlands 2014*, OECD Publishing, Paris. DOI: http://dx.doi.org/10.1787/9789264209527-en.

World Economic Forum (2013), *The Global Competitiveness Report 2013-2014: Full data Edition*, Geneva 2013. http://reports.weforum.org/the-global-competitiveness-report-2013-2014/#=.

Chapter 4

General incentives for investments in the Netherlands

This chapter reviews general incentives in the Netherlands for investments by firms, including farms, input suppliers, and food companies. It examines basic conditions for investments established by the overall regulatory environment; Trade and investment policy, which influences the flow of goods, capital, technology, knowledge and people needed to innovate; and access to credit needed to innovate. The general fiscal policy and the treatment of agriculture are then examined. The chapter is mainly based on the latest OECD Economic Survey of the Netherlands (OECD, 2014b). Specific obstacles and incentives for investment in the agricultural sector are dealt with in later chapters of this report.

The statistical data for Israel are supplied by and under the responsibility of the relevant Israeli authorities. The use of such data by the OECD is without prejudice to the status of the Golan Heights, East Jerusalem and Israeli settlements in the West Bank under the terms of international law.

Regulatory environment

The overall regulatory environment establishes basic conditions within which all firms, including farms, input suppliers, and food companies, operate and make investment decisions. Competitive conditions in domestic markets, including low barriers to entry and exit, can encourage innovation and productivity growth. Regulations may also enable or impede knowledge and technology transfer directly, contributing to more or less innovation. This section focuses on national regulations governing competition and entrepreneurship, natural resource management, and agricultural and food products and processes. It also indicates the scope of provincial regulations and provides some examples.

Regulatory environment for entrepreneurship

Dutch regulations for entrepreneurships are incorporated in the overarching frame of EU legislation on enterprises, which includes competition policy.

Figure 4.1. OECD Integrated Product Market Regulation (PMR) Indicator, 1998, 2008, 2013

Scale from 0 (least) to 6 (most) restrictive

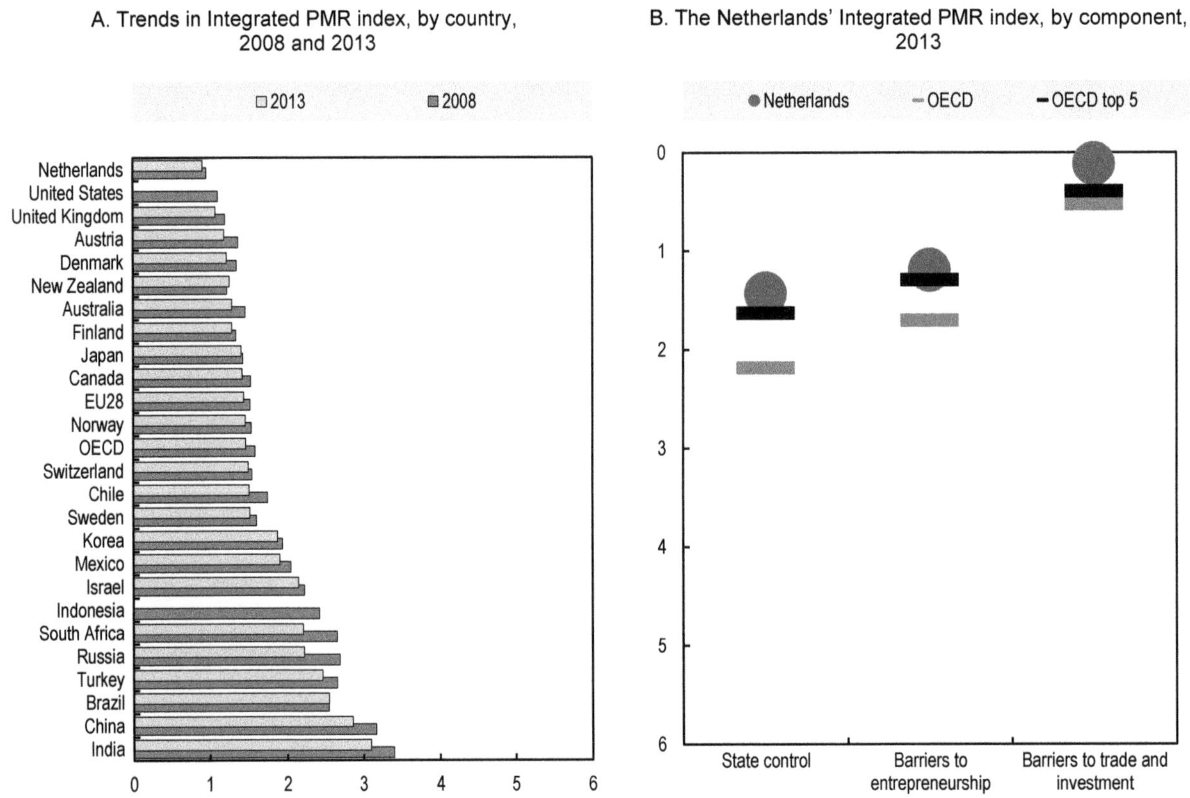

OECD top 5 refers to the average of the scores for the top five performers among OECD countries (Netherlands, United Kingdom, United States, Austria and Denmark), with US data referring to 2008.

Indices for EU28 and OECD are the simple average of member-country indices.

OECD Product Market Regulation (PMR) indicators measure key regulations in the areas of state control, barriers to entrepreneurship, and barriers to trade and investment.

Source: OECD Product Market Regulation Database, 2014. www.oecd.org/economy/pmr.

StatLink http://dx.doi.org/10.1787/888933250897

To lower administrative burdens, which involve comparatively higher costs for SMEs than for large firms, the Dutch government has rationalised its business support network. It has set targets to reduce administrative burdens and compliance costs for enterprises and improve transparency and provision of public services. The Netherlands Enterprise Agency (RVO), established in 2014 from a merger of agencies, offers help to facilitate compliance with laws and regulations (OECD, 2014b). A single desk platform (*ondernemersplein*)[1] was launched in 2014 to provide information and advice to entrepreneurs for free, as is the case for the RVO.

According to the World Bank's "Ease of doing business" the Netherlands has the most competition-friendly regulatory environment among the OECD countries, with the lowest overall score for regulation of total economy.

According to OECD Product Market Regulation (PMR) indicators, which measure the degree to which countries' regulatory frameworks promote or inhibit competition, regulations in the Netherlands have become less restrictive overall in the last 15 years, a trend found in many OECD countries (Figures 4.1 and 4.3A). The Netherlands has now the least restrictive PMRs of all OECD countries and emerging economies, below the United States and the United Kingdom. It is also the least restrictive of OECD countries for state control and barriers to trade and investment, and the third for barriers to entrepreneurship after New Zealand and the Slovak Republic, and thus figures among the OECD top 5.

Among barriers to entrepreneurship, the Netherlands is the least restrictive OECD country regarding administrative burden on start-ups and is among the OECD top 5 in terms of regulatory protection of incumbents (Figure 4.2). With its score on the complexity of regulatory procedures, the Netherlands does not make the top 5 of least restrictive OECD countries but is well above the OECD average. Figure 4.3B also shows specific areas where further efforts would be required to place the Netherlands among the OECD top 5 performers.

Barriers to entrepreneurship have fallen significantly over the last 15 years (Figure 4.3, Panel A), but there are still areas where they are higher than the average of OECD top 5 performers, in particular access to licences and permits (Panel B).

Figure 4.2. Barriers to entrepreneurship indicator, by regulatory area, 2013

Scale from 0 (least) to 6 (most) restrictive

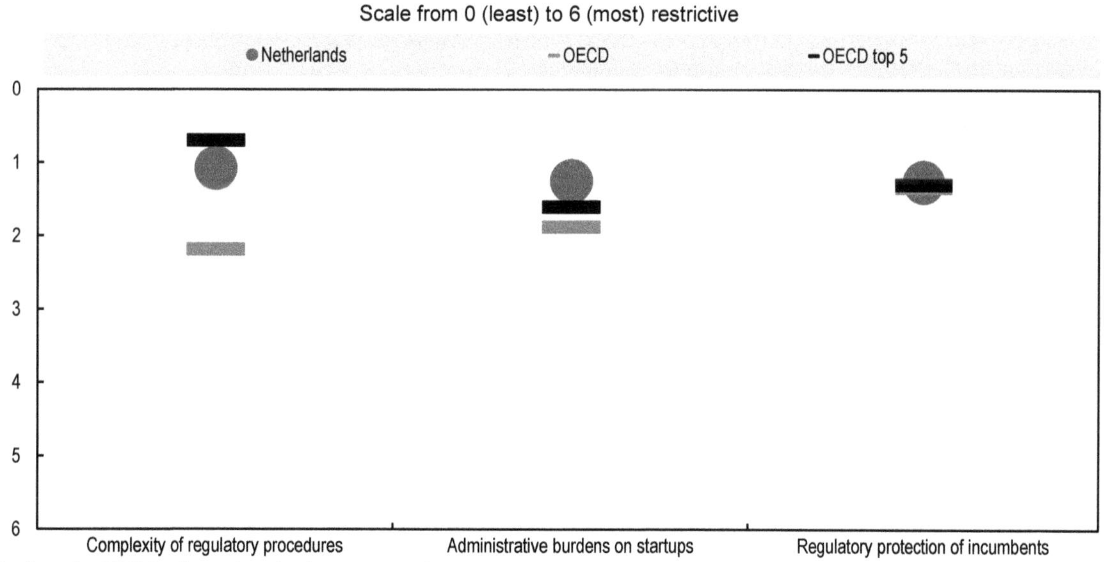

Indices for OECD all are the simple average of member-country indices.

OECD top 5 refers to the average of the scores for the top five performers among OECD countries (Slovak Republic, New Zealand, Netherlands, Italy and United States), with US data referring to 2008.

Source: OECD Product Market Regulation Database, 2014, www.oecd.org/economy/pmr.

StatLink http://dx.doi.org/10.1787/888933250909

Figure 4.3. Product market regulation (PMR): Barriers to entrepreneurship.

Index scale from 0 (least restrictive) to 6 (most restrictive)

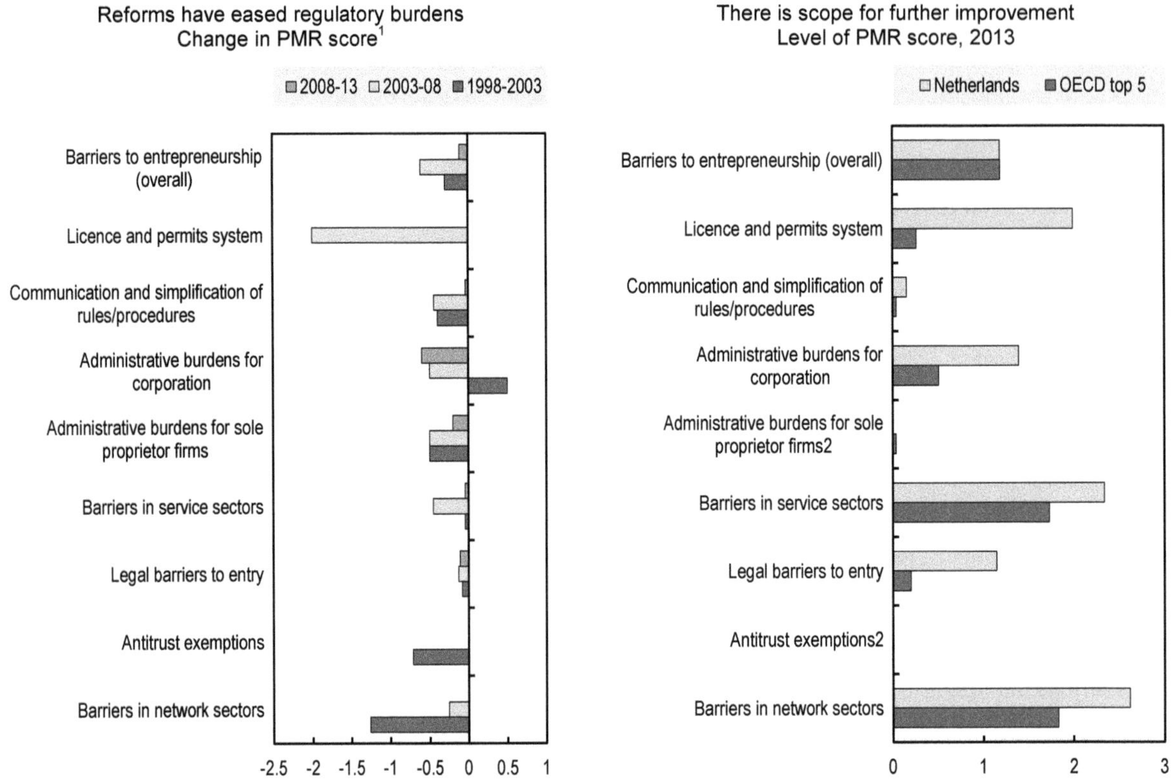

1. There was no change in the PMR score for licence and permits system in 1998-2003 and 2008-13, neither for antitrust exemptions in 2003-08 and 2008-13, nor for barriers in network sectors in 2008-13.

2. For administrative burdens for sole proprietor firms the PMR score of the Netherlands is zero (i.e. least restrictive). For antitrust exemptions the PMR scores are zero.

Source: I. Koske et al. (2014), "The 2013 Update of the OECD Product Market Regulation Indicators: Policy Insights for OECD and non-OECD Countries"; OECD Product Market Regulation Database, 2014. www.oecd.org/economy/pmr.

StatLink http://dx.doi.org/10.1787/888933250918

Regulations on natural resources

Regulations on natural resources are central to ensuring their long term sustainable use. In large part, they influence access to land, water and biodiversity resources and the impact that food and agricultural production systems have on those resources. The Dutch government is placing increasing emphasis on this area, as illustrated by the Government Programme for Natural Capital (Uitvoeringsagenda *Natuurlijk Kapital*), which aims to promote the resilience of ecosystems and ecosystem services by 2020, and includes a set of actions in a wide array of sectors, including agriculture (Rijksoverheid, 2015).

General regulations governing access to natural resources such as water and biodiversity

At the economy-wide level, the Netherlands is generally considered as having a high level of environmental regulation, especially in comparison with other OECD countries. This is illustrated by the Environmental Policy Stringency (EPS) index recently developed by OECD (Botta and Kozluk, 2014; Figure 4.4). The Netherlands ranks among the highest EPS index among the OECD countries considered, with an average EPS one-third above the OECD average. Moreover, between 1990-95 and 2012, there has been a four-fold increase in EPS in the Netherlands. These results are in line with other available indicators of environmental policy stringency (Botta and Kozluk, 2014).

Such a large increase in the stringency of environmental regulations does not seem to have impeded economic growth in the country over time. Botta and Kozluk (2014) suggest that, as far as OECD countries are concerned, there is no empirical relation between the change in environmental policy stringency and the trend of Multifactor productivity, except in the short run (less than five years). Results even tend to show the contrary in the long run, i.e. strengthening environmental regulations may be statistically associated with increases in productivity growth.

Figure 4.4. Stringency of environmental policy, 1990

Index scale from 0 (least restrictive) to 6 (most stringent)

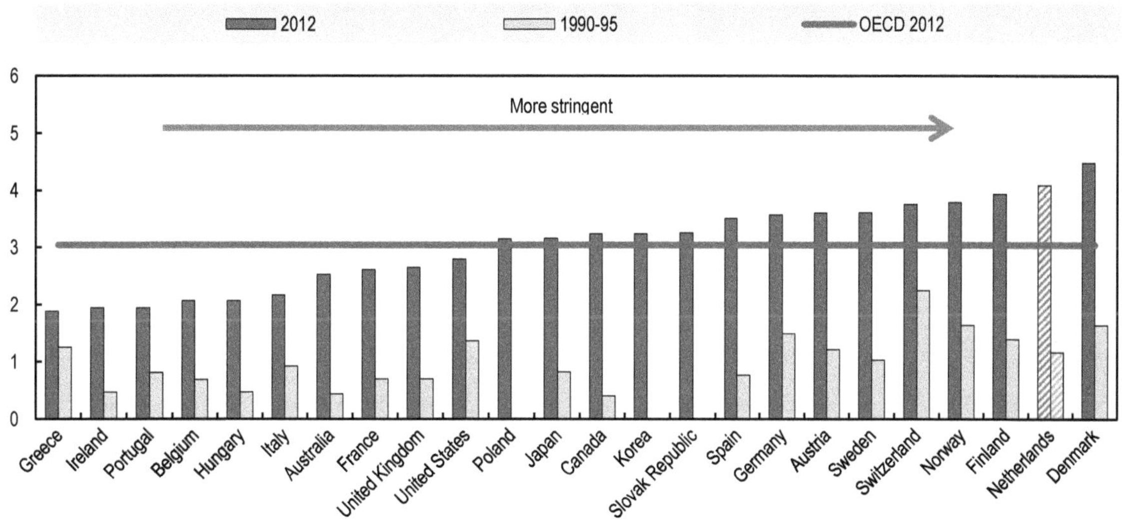

Source: Botta, E. and T. Koźluk (2014), "Measuring Environmental Policy Stringency in OECD Countries: A Composite Index Approach", *OECD Economics Department Working Papers*, http://dx.doi.org/10.1787/5jxrjnc45gvg-en.

StatLink http://dx.doi.org/10.1787/888933250925

To date, there is no equivalent EPS indicator for the agriculture sector, although some studies have estimated on a case-by-case basis the cost of environmental regulation for the farming sector in some areas, such as nutrients. Hence there is lack of evidence for the Dutch agriculture sector on a possible disconnection, or even positive effect of environmental regulation on productivity, as well as on innovation. Nevertheless, despite stricter environmental regulations, total factor productivity in agriculture is still increasing.

Environmental regulations

Environmental regulation in the Netherlands takes place in the framework of the Environmental Management Act (*Wet Milieubeheer*) and the Decree of general rules governing the environmental management of sites (*Activiteitenbesluit*). Most environmental regulations affecting agriculture in the Netherlands are incorporated in the overarching frame of EU regulations. EU environmental policy frameworks include notably:

- The Nitrate Directive.
- The Water Framework Directive.
- The Birds and Habitat Directives.
- The Crop Protection Policy, which has a strong European base. Both the authorisation of plant protection and use is aimed at reducing the risks to humans, animals and the environment.
- The EU Biodiversity Strategy to 2020 and (to lesser extent) the EU Forest Strategy.

- The Marine Framework Directive.

Given the European frameworks, there are policy themes with national priorities. In Dutch society there are serious concerns about how producers in the primary sector are dealing with the environmental pressures associated with land-intensive livestock production. Companies in the agricultural sector must also meet requirements in areas such as animal welfare, antibiotics, energy use, and environmental emissions. In addition to regulatory requirements, the producers (organisations) are also influenced through financial incentives, voluntary agreements, etc. In addition, there are policies to promote sustainable consumption and to prevent food wastage.

Nutrients

The European Nitrates Directive and the Water Framework Directive are setting standards on the amount of nitrate and nitrogen in groundwater and surface water. These standards limit the amount of manure that can be put on the land, and thus represent a potential constraint for productivity growth in the sector.

The Dutch policy regarding nutrient pollution and agriculture started to develop in the 1980s, and followed four major phases:

- 1984-87: Stabilization of livestock production.
- 1987-98: Progressive development of manure application limits and manure quota systems.
- 1998-2005: The Mineral Accounting System, which targeted farm level nutrient surplus.

2005-present: the current policy is based on application norms limiting the manure that can be sustainably used for different crops, animal production rights for poultry and pigs limiting manure production, and mandatory processing and (or) export of excess manure with the aim to limit the total of manure available for application (OECD, 2008). There is currently a derogation regarding the 170 of nitrogen per ha limit until 2014, for which an extension has been granted by the European Commission under the Nitrate Directive 5th Action Plan 2014-17.

The evolution of the Dutch nutrient policy in agriculture is interesting from the point of view of innovation, both organisational and technical. Since it started in the 1980s, the Dutch agricultural nutrient policy gradually moved from a regulatory, command-and-control approach to more market-based instruments, such as tradable quotas systems and the Mineral Accounting System. The policy also moved towards targeting *emissions* of pollutants (farm-level nutrient surplus), *rather than proxies* such as farm practices or inputs like fertilisers or manure applications. Conventional economic analysis suggests that such evolutions are more likely to foster innovation in terms of pollution reduction practices and techniques. Indeed, command-and-control approaches based on farm practices leave less freedom to farmers to adapt their production mixes or to innovate. Such changes in policy design could themselves be considered as organizational innovations, requiring a combination of information tools, accounting approaches to nutrients, monitoring systems, etc.

As shown in Chapter 2, nutrient balances have indeed decreased significantly since 1990, and the rate of decrease has been especially high under the period covered by the MINAS system (1998-2005). There is some evidence that the MINAS programme triggered a wave of innovative investments, but also fostered structural change in the sector, with a significant number of quota sellers quitting the farm sector (Wossink, 2003). Such structural change could have had indirect positive impacts on innovation, although there is a lack of evidence in this area. More directly, Wossink (2003) mentions that in the swine industry improvements of nutritional efficiency of diets and modified feeding regimes allow improving the nutrient efficiency of pig production.

In spite of encouraging reductions in nutrient surplus, the Dutch MINAS system was abandoned in 2005. Several reasons have been discussed for abandoning the system. A first set of issues concerned the design and practical implementation of the MINAS policy, including: uncertainty about the calculation of the farm-level nutrient surplus; significant transaction costs; complexity of the trading system (non-pecuniary costs to farmers). A key external driver was the decision of the European Court of Justice that the MINAS system was not in accordance with the Nitrate Directive. In the current system based on application norms, some problems remain in terms of policy design, especially regarding possible deviations between calculations and measurements of nutrient budgets, which can reach 20% in certain cases (LEI, 2015).

Pesticides and plant protection

The declining pesticide use in the Netherlands can be linked to the implementation in 1998 of the Policy Document on Sustainable Crop Protection to 2010, the objective of which was to limit risks to health, the environment (in particular surface and drinking water), and food safety at acceptable limits, while ensuring it does not reduce economic prospects for the agriculture sector. Reducing contamination of surface waters was particularly important in the light of the Water Framework Directive, which sets chemical and ecological quality standards for surface waters (Milieu-en Natuurplanbureau, 2015). Substantial improvements in terms of ecological quality of waters, drinking water quality, food safety and the safety of working conditions have been observed during the period 1998-2010.[2] Despite such positive trends, the initial policy objectives have not been reached yet (Table 4.1). Distance to targets to comply with the Water Framework Directive vary among policy objectives; for instance, the reduction of pressures from pesticides and herbicides to surface waters have been reduced by 85% between 1998 and 2010; compared to a target of 95%. On the other hand, the implementation of the crop protection policy has not reduced economic prospects for agriculture and horticulture in the Netherlands.

Decreasing environmental pressures arising from crop protection can be attributed to the increasing use of low-drift spraying techniques and the development of less-polluting pesticides (PBL, 2012). In that sense, most of the observed improvements come from a combination of adoption of best available technologies and of eco-innovation in the pesticide market. Efforts to ensure the sustainability of crop protection (and stimulate integrated pest management) are further pursued under the National Action Plan on Sustainable Plant Protection over the period 2013-18, as required by the Directive on the Sustainable Use of Pesticides (Directive 2009/128/EC). A new type of approach is adopted: rather than targeting the environmental burden, the Second Memorandum on Sustainable Crop Protection (2013-23) aims to "reduce the number of violations of quality standards by 90%" (Berkhout et al., 2014).

Table 4.1. Trend in sustainable crop protection and targets achievement

Objective	Indicator	Trend policy term	Objective achieved?
Ecological quality	Ecological quality – surface waters	Cannot be determined	No
	Environmental pressures on surface waters due to agriculture	Large improvements	No Realised: -85% between 1998 and 2010 Target: -95% between 1998 and 2010
Drinking water quality	Problems related to drinking water quality	Large improvement is likely	No
Food safety	Exceedance of maximally permitted residue levels in food	Large improvements	Yes
Safe working conditions	Risk inventory and evaluation	Slight improvement	No
Maintaining economic prospects	Economic prospects in relation to this policy	Unchanged	Yes

Source: PBL (2012), *Evaluatie van de nota Duurzame gewasbescherming*, http://www.pbl.nl/publicaties/2012/evaluatie-van-de-nota-duurzame-gewasbescherming.

Another key change is that authorisation holders (pesticide manufacturers), rather than farmers, are accountable for exceeding standards in terms of water quality, with the underlying idea of creating a stewardship incentive to ensure that farmers make the appropriate usage of their phytosanitary products. So far this new approach can be considered as an interesting innovation in terms of policy design. The impact of this change on economic and environmental performances, as well as in terms of dynamic incentives for technological innovation in this area has not been studied yet, but it would be interesting to conduct an *ex post* policy assessment. For the long term, the focus could be on investments in larger system innovations, and in less-polluting substances and non-chemical methods, such as biological control, which can be defined as the reduction in population densities of harmful organisms by exploiting one or more natural enemies of those organisms (PBL, 2012).

Regulation on land use and biodiversity

The Netherlands has a long tradition of land use planning (OECD, 2008)[3]. Many of these planning activities have focused on controlling water and creating land from wet delta lands. For 50 years, national land use plans aimed to maintain the separation between urban and rural areas, and strict zoning arrangements protected the agricultural land base from conversion (OECD, 2008). However, land use policies have over the last decade loosened this conceptual separation, and land use policies have become decentralised. The authority has been delegated to provinces, with the national government more likely to see its role as limited to deciding basic conditions in fields like land-use planning. At the same time, the focus has shifted from imposing restrictions to promoting developments. The concept of developmental land use planning has been introduced, stressing the importance of area development carried out by public, private and civil society actors. This introduces competition for land use between nature conservation, agriculture, and other economic activities.

The role of agriculture in contributing to the connectivity between natural areas, through areas dedicated to lower intensive farming and green infrastructure is supported by the Natura 2000 policy, and the associated Dutch National Ecological Network, which includes agricultural land under nature-friendly management; and related subsidies for agricultural nature management. Innovation in ecological sciences and application would be key in this domain, to find through combined ecological-economic research and development more cost-efficient approaches to reduce landscape fragmentation through changes in agricultural practices and land uses. In this area, it would be worth investigating in more detail the role that agri-environmental measures could play.

Climate change policies

Despite the fact that the nature and degree of the projected impacts of climate change in the Netherlands remain uncertain, especially for agriculture, the country has undertaken several policy initiatives in the sense of adaptation. Such initiatives include the *Delta Programme* aiming at redesigning flood risk management, including agricultural land areas; the assessment with OECD of water governance systems and how to improve them (OECD, 2014c); and the possibility to further develop resilience through more ecosystem-friendly approaches, such as recommended recently by PBL (PBL, 2013). The risk of livestock disease may increase with climate change, prompting renewed interest in this area. The Netherlands has notably contributed to a recent OECD workshop on the management of livestock disease (Bergevoet and van Asseldonk, 2013).[4]

Regulations on products and processes

Regulations on products and processes aim to protect the environment and human, animal and plant health and can also impact natural resource use. In developing an appropriate sanitary and phytosanitary (SPS) regulatory environment, including implementation provisions, experience has shown that technology neutral, science-based approaches are most effective in diffusing innovation and least market distorting provided that care is taken to ensure agricultural specificities and societal choices are taken into account. There is also evidence that good product market regulation is associated with increased inflows of foreign direct investment and thus technology spill-overs. Environmental and health related regulations could boost innovation by building consumer and societal trust in the safety and sustainability of new products or processes, but unnecessary or disproportionate regulations can stifle innovation and technological developments.

Regulations on product and processes are mainly determined at the EU level and implementation is at the national level. Furthermore, the Dutch government actively encourages sustainable food production in the Agenda for Sustainable Food by making agreements with companies about for example waste reduction (Dutch Alliance for Sustainable Food) and making livestock farming more animal friendly (sustainable livestock farming implementing agenda).

EU regulatory approach on food safety

Examples of regulatory practices in the European Union are given in Box 4.1. A variety of innovative approaches can help reduce the regulatory cost burden for governments. These include use of public private partnerships based on "best practices" in the way the SPS regulatory framework is managed, including the interface between private voluntary standards and compulsory compliance regulation (OECD, 2013).

For the food processing industry, EU regulation can be complex and open to interpretation. This can be an issue, in particular for SMEs. Another issue is the length of time for the approval of new products on the market, which is typically 4 or 5 years compared to 6 months in the United States. But this does not seem to have prevented Dutch and other EU countries' industries from being very innovative.

National institutions and practices for the implementation of standards and evaluation procedures

To ensure food safety and food quality, the government sets rules, monitors compliance and sanctions where necessary. Another purpose of the government is to secure plant and animal health. For this the government has special guidelines, maintains a knowledge infrastructure and controls the products that are imported in the European Union. The government coordinates the control of animal diseases and acts on suspicions of animal diseases. In this way the government contributes to the prevention of outbreaks of infectious plant and animal diseases, to food safety and public health, to the international trade position of the sector and to the conservation of flora and fauna.

The Netherlands Food and Consumer Product Safety Authority (NVWA) is in charge of controlling the sector for food safety and animal welfare. It is made of independent experts and its activities are performed under the supervision of the Ministry of Economic Affairs with close links to the Ministry of Health, Welfare and Sport. The industry pays for most of the cost of inspection and control, with little government money, but the government has full responsibility for inspection and supervision.[5]

Box 4.1. EU regulatory practices

Smarter regulation in the European Union

Smarter regulations aim to simplify existing EU legislation in order to spur innovation and reduce the administrative burden for operators. Independent evaluations have been commissioned on several legislative areas including Genetically Modified Organisms (GMO), animal health, plant health and seeds. Impact assessment is now required for any regulatory proposal to improve the quality of proposals, ensure consistency between Community policies, and contribute to sustainable development. In terms of innovation, impact assessment takes the following questions into consideration:

- Does the option stimulate or hinder R&D?
- Does it facilitate the introduction or dissemination of new production methods?
- Does it affect IPRs, including patents, trademarks, copyrights and other "know-how" rights?
- Does it promote or limit academic or industrial research?
- Does it promote greater productivity or resource efficiency?

Source: Gerlitz (2012).

EU legislative framework for ensuring GM food and feed safety based on Directive 2001/18/EC

The European Food Safety Authority (EFSA) is the agency responsible for the risk assessment regarding food and feed safety. In close collaboration with national authorities and in open consultation with its stakeholders, EFSA provides independent scientific advice and clear communication on existing and emerging risks. EFSA risk assessment procedures are based on international standards and are often defined in the scientific arena as the most comprehensive risk assessment procedures in the world. The three typical steps of the EFSA GM food/feed risk assessment process are: 1) Molecular characterisation; 2) Compositional analysis; 3) Food and feed safety analysis and risk evaluation; 4) environmental impact analysis.

The risk management phase is managed by the European Commission and member states. In order to obtain an authorisation for the production of GM food products, the interested parties have to submit an application to the competent national authorities, which has to acknowledge the receipt of the application and inform the EFSA without delay. Applications are sent to the European Commission and to the member states, who are consulted on the application over a three month period. EFSA must provide its opinion within six months of receiving the application. However, if additional data is requested during the scientific assessment the time limit is extended. The services of the Commission have to take due account of the comments of the public (within one month after the EFSA opinion) and submit a proposal agreed by the different departments of the Commission (inter-services consultation) to a committee composed of representatives of the member states and go through an examination procedure. When a qualified majority occurs in the Committee, the decision is adopted, published in the Official Journal of the European Union and included in the above-referred GMO register. Otherwise, the Commission must refer the issue to the Appeal Committee, which will have a two-month timeframe to adopt a decision. Adoption is possible in the absence of a decision.

Authorisations, when granted, are valid for ten years and are renewable for ten years each time. However, the decision can be reviewed and even withdrawn at any time if new elements occur that would justify such an intervention. In other words, the Commission with the fundamental scientific advice of EFSA maintains a substantial supervision power. Finally, all authorized products are entered in the EU register, which contains all relevant details and information.

Recent developments

Amending Directive 2001/18/EC, Directive (EU) 2015/412 of 11 March 2015 gives member states flexibility to decide on the cultivation of genetically modified crops on part or all of their territory, under certain conditions, at two distinct points in time:

- During the authorisation procedure: a Member State can ask to amend the geographical scope of the application to ensure that its territory will not be covered by the EU authorization.
- After a GMO has been authorised: a Member State may prohibit or restrict the cultivation of the crop based on grounds related amongst others to environmental or agricultural policy objectives, or other compelling grounds such as town and country-planning, land use, socio-economic impacts, co-existence and public policy. These grounds shall in no case conflict with the environmental risk assessment carried out in this context.

In addition, as from April 2017, member states in which GMOs are cultivated shall take appropriate measures in border areas of their territory with the aim of avoiding possible cross-border contamination into neighbouring member states in which the cultivation of those GMOs is prohibited.*Source:* Valletta (2010) and Directive (EU) 2015/412 of the European Parliament and of the Council of 11 March 2015. http://eur-lex.europa.eu/legal-content/EN/TXT/PDF/?uri=CELEX:32015L0412&from=EN .

Trade and investment policy

Trade can facilitate the flow of goods, capital, technology, knowledge and people needed to innovate. Openness to trade and capital flows is conducive to innovation as it provides a larger market for innovators, reinforces competition, increases access to new technologies, ideas and processes, including from Foreign Direct Investment (FDI) and related technological spill-overs, and facilitates cross-country collaboration. Trade and investment openness can influence innovation throughout the food supply chain, from input suppliers to food service and retail firms. Input and output markets that operate effectively can foster productivity growth and more environmentally sustainable production (OECD, 2014).

Barriers to trade and investment

The Netherlands applies EU common tariffs. Industrial tariffs are lower than agricultural tariffs (simple average of applied most-favoured-nation (MFN) tariffs) (Figure 4.5). EU tariff for capital and intermediate goods are higher than in major OECD trade partners. Lower tariffs on intermediate goods would lower the cost of specialised inputs and machinery equipment and thus the competitiveness of the agri-food sector.

OECD PMR indicators also show that the Netherlands is the country with the lowest restrictions to trade and investment among OECD countries and emerging economies (Figures 4.2 and 4.6).

The Foreign Direct Investment (FDI) regulatory restrictiveness Index (FDI Index) measures statutory restrictions on foreign direct investment. Figure 4.7 shows that the Netherlands has a very open economy with hardly any restrictions on Foreign Direct Investment (FDI), far below the OECD average, and no restriction on investment in agriculture and food.

Figure 4.5. Tariffs for industrial and agricultural goods, 2011 or 2012

Simple average MFN applied tariff rates[1]

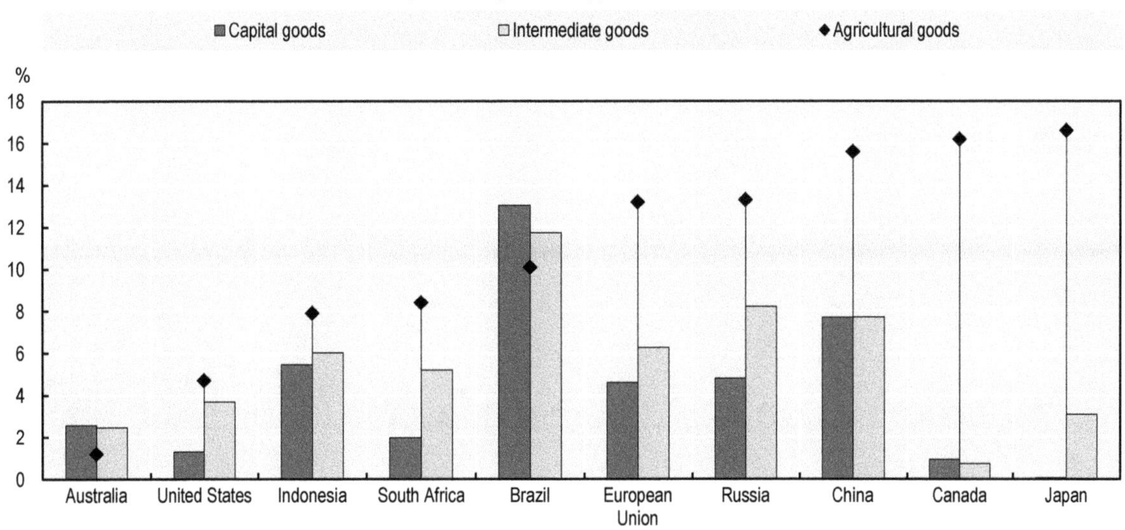

MFN: Most favoured Nation.

1. Tariff rates for agricultural products include both ad valorem duties and specific duties in ad valorem equivalent, while tariff rates for agricultural products only include *ad valorem* duties.

Source: UNCTAD Trade Analysis Information System (TRAINS) (for non-agricultural products) and World Tariff Profiles, 2013 (for agricultural products).

StatLink http://dx.doi.org/10.1787/888933250930

Figure 4.6. Index of regulatory restrictions to trade and investment, 2013

Scale from 0 (least) to 6 (most) restrictive

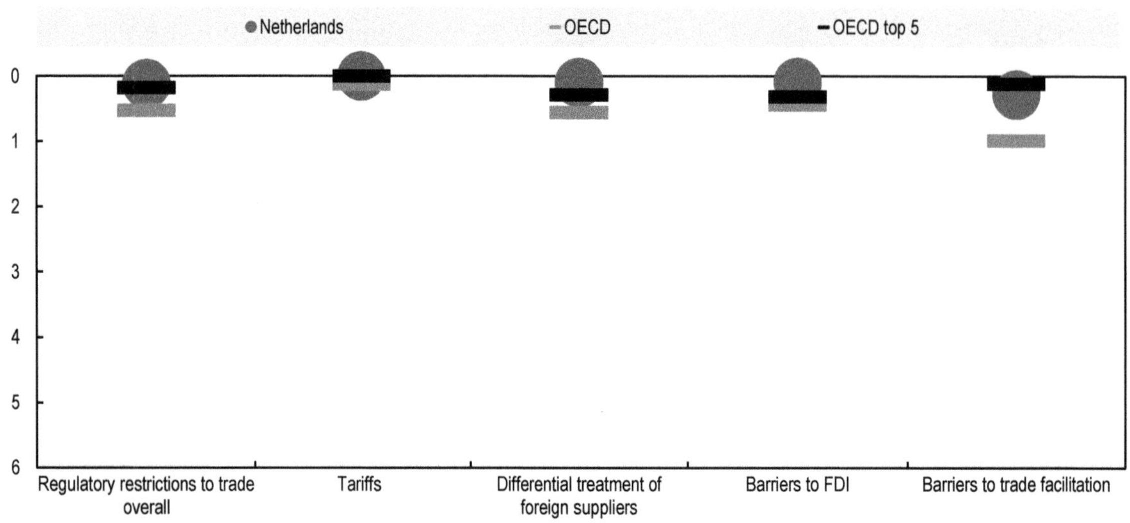

Barriers to trade facilitation refer to the extent to which the country uses internationally harmonised standards and certification procedures, and Mutual Recognition Agreements (MRAs) with at least one other country.

OECD top 5 refers to the average of the scores for the top five performers among OECD countries (Netherlands, Belgium, Australia, United Kingdom and Finland).

Source: OECD Product Market Regulation Database, 2014. www.oecd.org/economy/pmr.

StatLink http://dx.doi.org/10.1787/888933250949

Figure 4.7. OECD FDI Regulatory Restrictiveness index, by sector, 2013

Scale from 0 (least) to 1 (most) restrictive

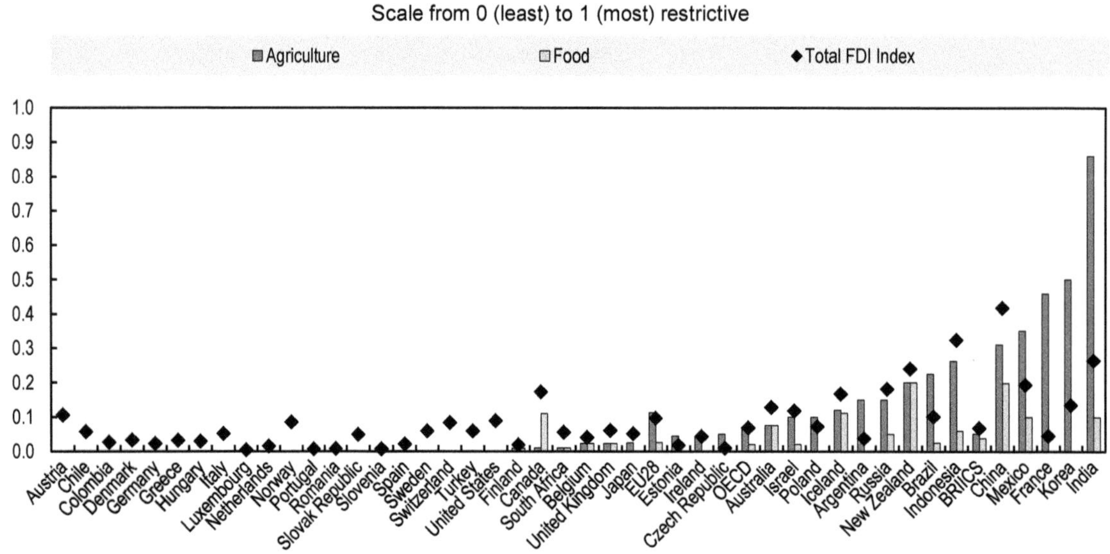

Indices for OECD are the simple average of member-country indices.

Four types of measures are covered by the FDI Restrictiveness Index: 1) foreign equity restrictions, 2) screening and prior approval requirements, 3) rules for key personnel, and 4) other restrictions on the operation of foreign enterprises.

Source: OECD Investment Statistics. http://www.oecd.org/investment/fdiindex.htm.

StatLink http://dx.doi.org/10.1787/888933250956

FDI positions and flows

FDI flows, which concern mainly the food manufacturing sector, are a means to transfer technologies and know-how across partner countries. In the last two decades agri-food FDI positions have increased over time despite some slowing down of flows in the mid-2000s and early 2010 (Figure 4.8). The Netherlands is generally a net exporter of agri-food FDI and embedded knowledge, but the export surplus has become increasingly variable since the mid-2000s. The Netherlands also attracts foreign FDI, including in agri-food, through regional structures such as Food and Seed Valleys (Box 5.1)

Figure 4.8. Netherlands' Foreign Direct Investment in food and agriculture, 1990-2012

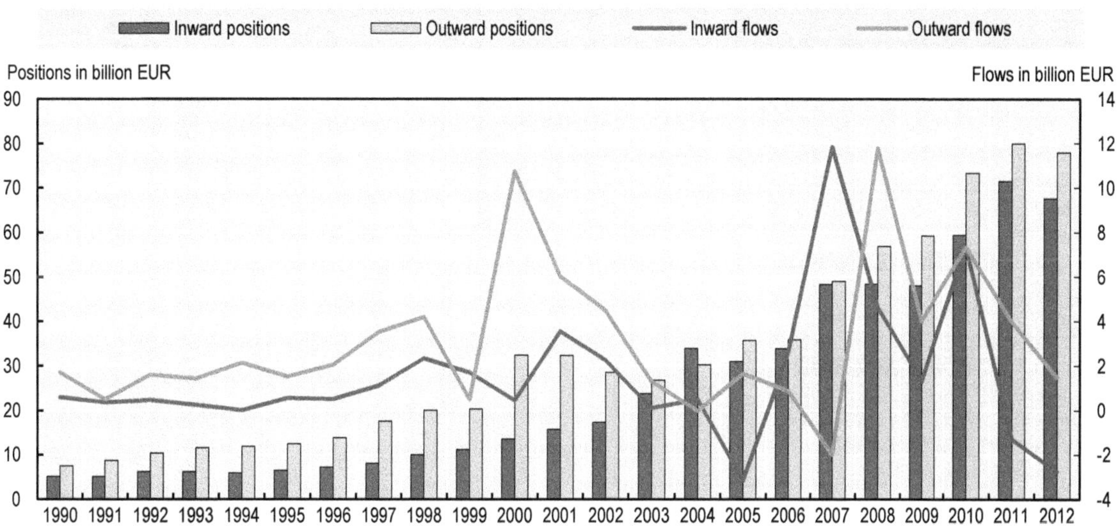

Source: OECD Investment Statistics. http://www.oecd.org/investment/fdiindex.htm.

StatLink http://dx.doi.org/10.1787/888933250961

Finance policy

Efficient financial services are one way to enable balanced development of any economy and society. Policies that improve the functioning of financial markets can facilitate productivity enhancing investments in agriculture. Low-cost loans and venture capital[6] can also be an important source of funding for innovative firms with high growth sectors potential. Business angels[7] also play an important role in financing early stages of innovation (OECD, 2014a).

Financial market development

According to the WEF Global competitiveness indicator, the Netherlands scores slightly higher than the OECD average in terms of financial market development (Figure 4.9). It obtains high scores and figures among the OECD top 5 performers for the availability and affordability of financial services, but lower scores than the OECD average for the availability of loans and venture capital, and soundness of banks. This reflects the financial crisis of 2008, which widely affected the European banking system. Moreover, the strength of legal right index is lower than the OECD average, indicating that collateral and bankruptcy laws in the Netherlands are less protective of the rights of borrowers and lenders, and thus do not facilitate lending to the same extent. This is linked to the difficulty of access to credit.

Figure 4.9. Global Competitiveness Index: Financial market development, 2013-14

Scale 1 to 7 (best)

A. Total index of financial market development, by country

B. Netherlands' index of financial market developments, by component

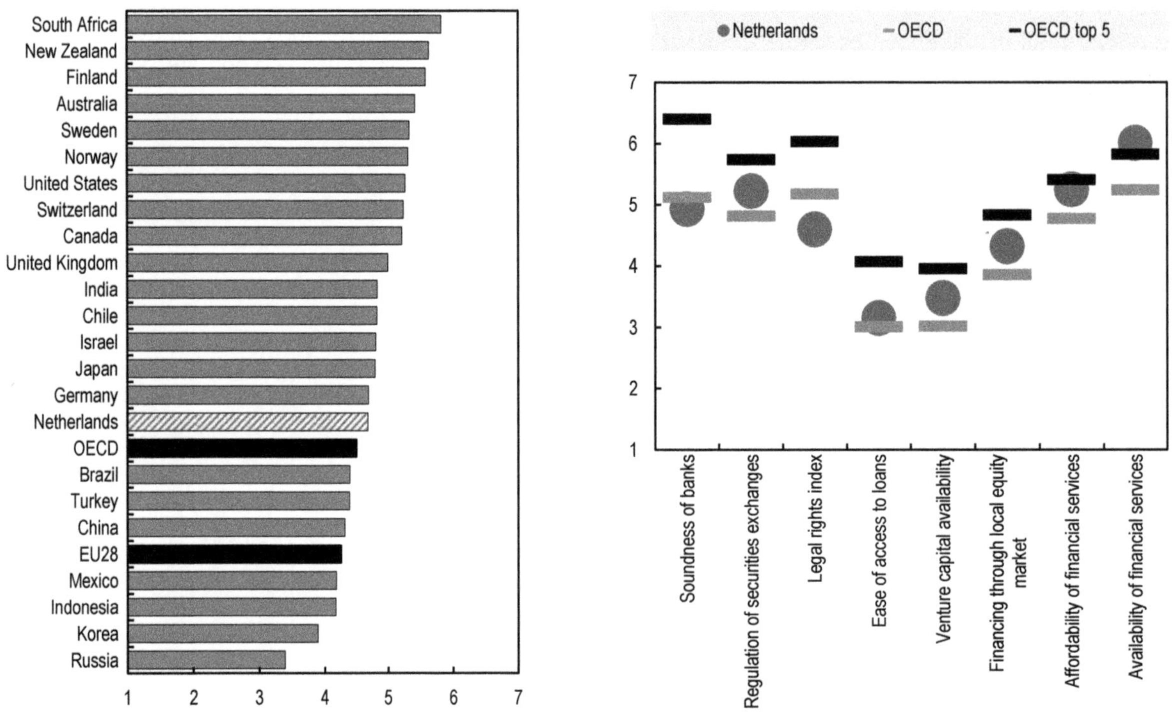

Indices for EU28 and OECD are the simple average of member-country indices.

Top 5 refers to the average of the scores for the top 5 performers among OECD countries (New Zealand, Finland, Australia, Sweden and Norway).

The Legal rights index is scored on a scale from 1 to 10 based on calculations by the WEF from the World Bank–International Finance Corporation's Doing Business 2013.

Source: World Economic Forum (2013), *The Global Competitiveness Report 2013-2014: Full data Edition*, http://reports.weforum.org/the-global-competitiveness-report-2013-2014/#.

StatLink http://dx.doi.org/10.1787/888933250975

The Dutch banking sector is large compare to the economic size of the country and has suffered major losses on international capital markets at the beginning of the financial crisis. Despite progress to strengthen bank capital, exposure to non-performing loans not covered by loan loss provisions is high. In addition, the dependence on international capital markets has increased remains large and the volatility of risk premia has increased (OECD, 2014b).

Dutch banks have been reporting reductions in loan demand since the beginning of the crisis in 2008, but they have also been rationing credit as lending standards have been tightened, mainly through stricter collateral requirements reported by SMEs. In turn, tight credit standards have been weighing on business lending. According to bank lending surveys, costs related to the capitalisation of banks have had a lower impact on lending conditions than poor industry, firm and economic outlook. Nearly 20% of all surveyed Dutch SMEs reported obstacles for receiving a bank loan around mid-2013, one of the highest ratios in the euro area (Figure 4.10).

Figure 4.10. Bank lending constraints for SMEs[1]

Percentage of all respondents, April to September 2013

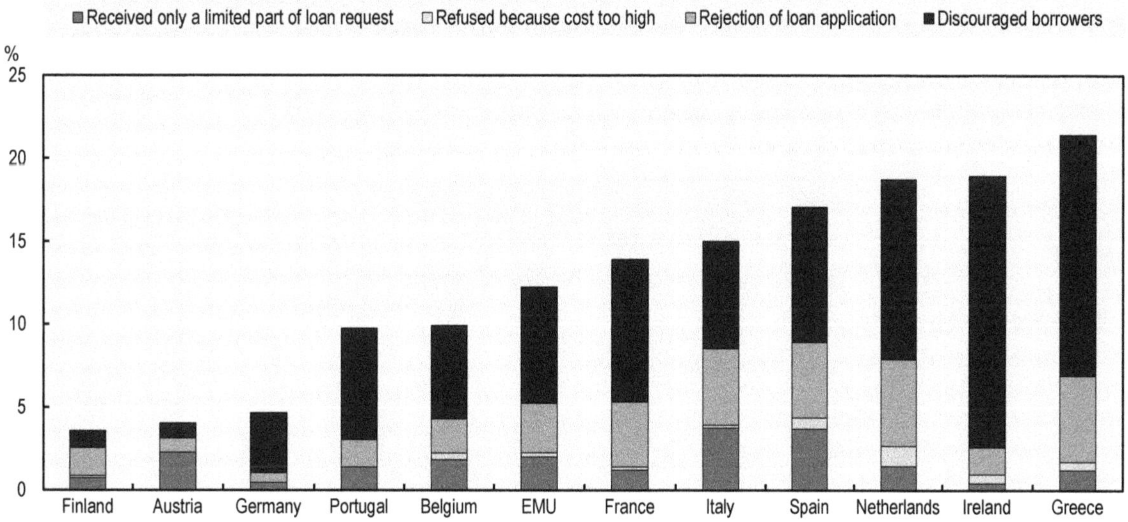

1. SMEs are defined as having 0-249 employees. EMU: European Monetary Union.

Source: ECB (2014), "Survey on the Access to Finance of SMEs", *Statistical Data Warehouse*, European Central Bank, March and DNB (2014), "Domestic MFI-statistics", *Statistics DNB*, De Nederlandsche Bank, March.

StatLink http://dx.doi.org/10.1787/888933250983

Another issue raised by companies in agri-food and other sectors is the need to find multiple sources of finance, and the complexity to navigate through all the options and programmes.

Access to finance (loan and loan guarantees)

According to a European Central Bank Survey for part of 2013, SMEs in the Netherlands have used external financing to a larger extent than their Eurozone counterparts (Figure 4.11). They have had access to grants and subsidised bank loans to a lesser extent, and have to use unsubsidised bank loans to a larger extent.

The recent scarcity of bank lending, combined with the limited role of venture capital in risk financing, have limited investments by businesses (OECD, 2014a). In response, the Dutch government has launched a number of programmes to ease access to finance. These include higher guarantees to banks for lending to SMEs and start-ups with little or no collateral, and the possibility to delay the repayment of loans benefitting from state guarantees.

Together with banks, the authorities started a microcredit institution, Qredits, in 2009. Public guarantees of equity stakes for venture capital investors and/or subordinated loans made by banks further eased small business finance. Other measures have aimed to stimulate direct public lending to new, fast-growing and innovative companies or to attract private investors (such as business angels) through public co-investments. More recently, public guarantees have been extended to non-bank institutions. The objective is to promote the development of credit unions or crowd funding, but also to entice pension funds and insurers into a planned SME financing fund and Netherlands Investment Institution.

The list of programmes that support business investment and their characteristics are presented in Table 7.4. They include the SME loan guarantee scheme (BMKB); Growth facility scheme (RG); Business loan guarantee scheme (GO), Netherlands Investment Institution (NII), and Microfinancing (by Qredits).

Figure 4.11. Sources of external financing of SMEs

Percentage of all respondents, April to September 2013[1]

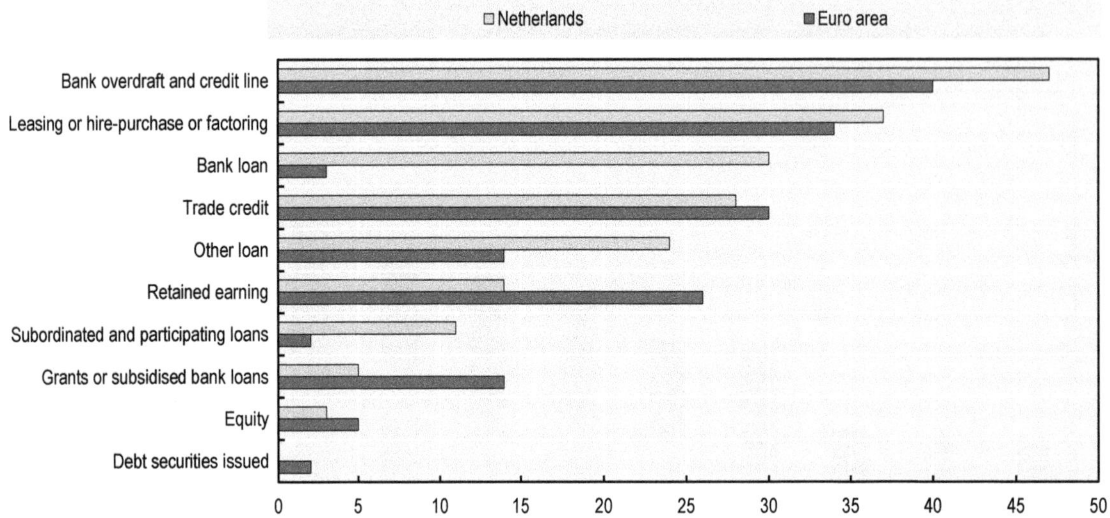

1. Figures refer to the following question: "Turning to the financing structure of your firm, to finance normal day-to-day business operations or more specific projects or investments, you can use internal funds and external financing. For each of the following sources of financing, could you please indicate whether you used them or not during the past six months?" Small and medium-sized enterprises (SMEs) are defined as having 0-249 employees. The category of subordinated and participating loans also includes preferred stocks and other similar instruments. The category of bank overdraft and credit line includes credit cards overdraft.

Source: ECB (2014), "Survey on the Access to Finance of SMEs", *Statisical Data Warehouse*, March, https://www.ecb.europa.eu/stats/money/surveys/sme/html/index.en.html .

StatLink http://dx.doi.org/10.1787/888933250995

In addition to the variety of credit support mechanisms available to businesses for various types of investment, the Dutch government has sought to ease the availability of credit for innovation-related investments. This has come as a response to the tight credit conditions since the beginning of the financial crisis in 2008, especially for SMEs. The instruments vary according to the target groups (small or larger firms) and the stage of financing or venture capital services provided.

Those programmes are also briefly described in Table 7.4: the SME+ Innovation Fund (MKB+) provides innovation credit, and supports private equity firms investing in early stage start-up companies (seed facility). Mechanisms are also used to stimulate innovation demand: The Small Business Innovation Research Programme (SBIR) and the Innovative Procurement Urgent (*Inkoop Innovatie Urgent*) provide project funding for public procurement responding to societal challenges, demand stimulation, or valorisation of public knowledge. Chapter 9 reviews all forms of support to innovation and discusses their relative contribution to general and agricultural innovation.

Over the years, the **Rabobank** has played an important role in the development of Dutch agriculture and horticulture, in the country and abroad. The Rabobank is now a multinational, co-operative group comprising independent local Rabobanks plus Rabobank Nederland, their umbrella organisation, and a number of specialist subsidiaries. It developed more than a century ago as a farmers' bank and it still holds an 85% to 90% market share in the farm sector in the Netherlands. Throughout the years, the company has also started targeting small and medium-sized companies, including agri-food companies. In 1987, the total outstanding loans in sectors other than agriculture exceeded those in the agricultural sector for the first time. By 2005 the agricultural credits amounted to some 8% of total outstanding credit.[8]

Figure 4.12. Rabobank credit to agricultural sectors in the Netherlands, 2005-14

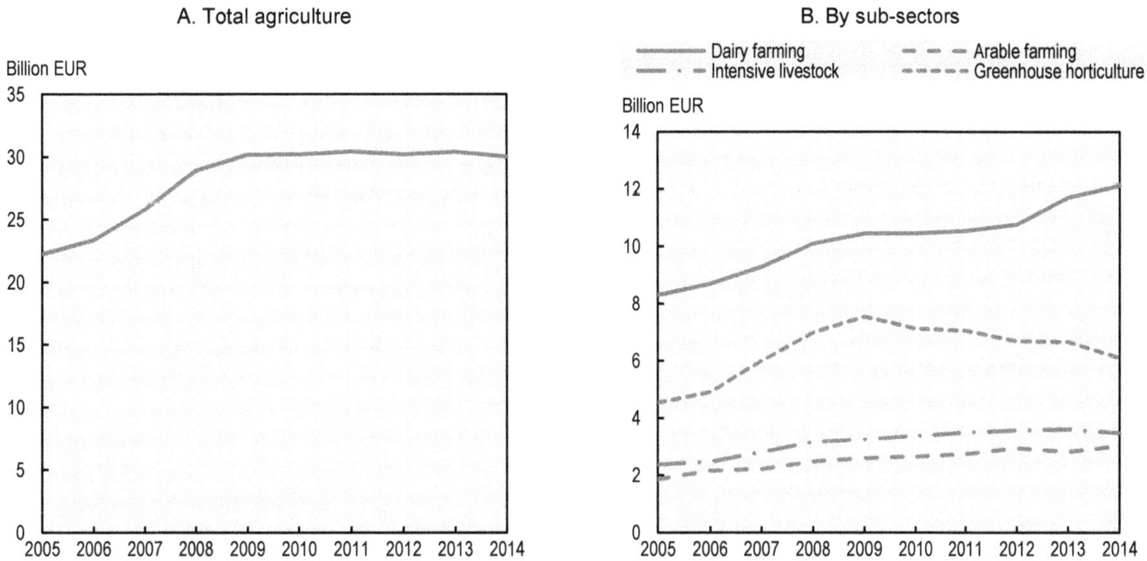

Intensive livestock = pig farming and poultry farming; greenhouse horticulture = vegetables, flowers and plants under glass.
Source: Rabobank.

The lending volume to Dutch agriculture from the Rabobank was approximately EUR 30 billion in 2014. While the volume of credit increased rapidly before 2009, it has remained fairly stable since (Figure 4.12.A). There are differences in the development of credits between agricultural sub-sectors. Dairy farming is the largest borrower of all sub-sectors over the period 2005-14 (Figure 4.12.B). The total volume of credit has increased by close to 50% over that period, in particular since 2013 as dairy farmers invested in preparation for the abolition of the EU milk production quota. The volume of credit in arable farming has gradually increased over the past 10 years. Better prices for grain and potatoes have supported income development in arable farming. This contrasts with income developments in greenhouse horticulture. In the period 2006-09 a lot of money was invested in greenhouses and the volume of credit from Rabobank increased substantially. In recent years, the greenhouse sector (especially for vegetables) has suffered from depressed incomes and high debts, and several companies have gone bankrupt. From 2009 to 2014 the total volume of credit in greenhouse horticulture has declined significantly. The outstanding volume of credits in intensive livestock has been stable during the last five years.

Tax policy

Tax policy affects innovation, productivity and sustainability in many ways: it affects the decision of firms and households to save or invest in physical and human capital, and thus the adoption of innovation; it raises government revenues, which can then finance public services, including those enabling innovation such as education and skills, R&D, and strategic infrastructure; it can also be used to provide direct incentives, for example preferential tax treatment to investments in private R&D or to young innovative companies. In addition to its economy-wide impacts, tax policy influences the conduct, structure and behaviour of farm, input suppliers and food companies. Farms and agri-food firms are generally subject to the same taxation regime as the rest of the economy. While they can enjoy some tax reduction arrangements, they are generally not specific to agriculture (OECD, 2005). Specific arrangements exist, however, for farm land, and reduced energy tax for greenhouses and full exemption for gas used for combined heat and power, which are expected to benefit the agricultural and horticulture sector.

Taxes on personal and corporate income

The average worker in the Netherlands faced a tax burden on labour income (tax wedge) of 36.9% in 2013 compared with the OECD average of 35.9%. The Netherlands was ranked 20 of the 34 OECD member countries in this respect. The tax burden for workers has, however, declined between 2009 and 2013.

Since 2011, incorporated small businesses are taxed at a preferential corporate tax rate of 20% up to EUR 200 000 of taxable profit, against a basic rate of 25%, which is at the OECD median (Figure 4.13).

Figure 4.13. Combined corporate income tax rate[1]

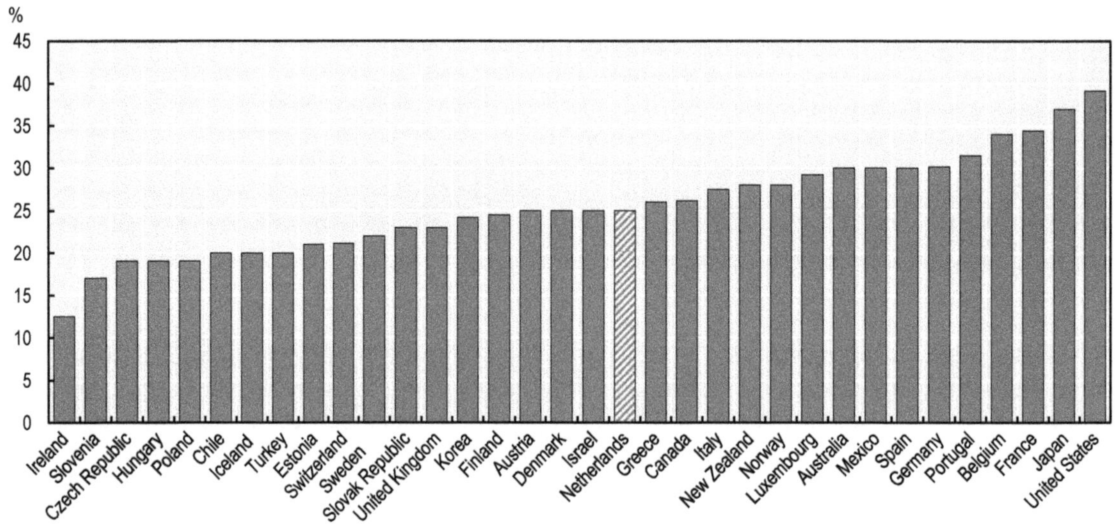

1. Basic combined central and sub-central (statutory) corporate income tax rate given by the adjusted central government rate plus the sub-central rate.

Netherlands: applies to taxable income over EUR 200 000.

Source: OECD Tax Database.

StatLink http://dx.doi.org/110.1787/888933251007

As all entrepreneurs, self-employed farmers can benefit from a general allowance to reduce their taxable income. This includes the self-employed allowance, a R&D allowance discussed later, a co-operation allowance and a cessation of business allowance. Entrepreneurs can make use of a fiscal reserve to provide for their old-age pension. Similarly, the income averaging facility, which allows averaging operating income over three years, applies to all business, including in agriculture (OECD, 2005). There is also the fiscal possibility to compensate for income losses one year backward and nine years forward.

Some specific income tax facilities are relevant to the agricultural sector:

- No tax is levied on specific woodland and nature subsidies.
- A specific tax allowance is given for investments in the environment, energy and small businesses. This allows a deduction of a portion of the environmental investment (not uniquely for the agricultural sector).
- A number of generic regulations provide tax facilities for certain green investments.

Taxes on property

A 2005 OECD report that reviewed tax exemptions for agriculture (OECD, 2005) also noted the differentiated treatment of land in the tax system, in particular agricultural land:

- Agricultural land, forestry, manors, nature and greenhouses are exempt from the **real estate tax** levied by municipalities.
- For the acquisition of farmland there are several exemptions to the **real property transfer tax** as well as for the acquisition of business real estate by children from the parents. There is a double taxation relief with regard to the inheritance and gift tax.
- The Agricultural Allowance *(Landbouwvrijstelling)* exempts capital gains on land from taxation. However, the law provides for two exceptions to this: 1) changes in value arising in the course of business (e.g. irrigation); and 2) non-agricultural changes in the value of land.

These are to facilitate farm transmission. The exception for land improvement under the agricultural allowance is expected to encourage investments in land improvement.

Motor fuel tax

Until the end of 2012, there was a tariff differentiation in the excise duty on mineral oils depending on the use of oil as fuel for vehicles on public roads (white diesel) or other uses such as heating and agricultural tractors (red diesel). The reason for this differentiation was to exclude vehicles that only use roads to a very limited extent. The tax deduction amounted to EUR 72.5 million in 2012 and has been abolished on 1 January 2013.

Energy tax in glasshouse horticulture

An Energy Tax applies to all large-scale users of energy (natural gas, other gases, electricity and certain mineral oils). The tariffs depend on the amount used, decreasing with greater use. There is a standard tax on energy (EB) and an additional charge (since 2013) for the Subsidy Scheme for Sustainable Energy (SDE +). This tax is also known as the Storage Sustainable Energy (ODE). Both EB and ODE have a decreasing tiered pricing structure: the levies are relatively lower for industrial energy users. There are special facilities for glasshouse horticulture because it has a small-scale company structure. A low tariff has been agreed for heating used to assist the growth of horticultural products. A reduced tax rate is applicable for the first 1 million m^3 of gas each year. This is related to agreements and covenants between the government and the horticulture sector with targets on CO_2 emissions, energy efficiency and increasing the share renewable energy. The sector will have to pay for exceeding the agreed CO_2 emissions target. Environmental taxation is levied on waste matter and the withdrawal of groundwater (in addition to energy and fuel covered above). The taxable person is the person who withdraws the groundwater, including farmers (OECD, 2005).

Tax incentives for R&D

Over the past two decades, OECD countries have increasingly employed tax incentives for R&D. The Netherlands was among the first countries to introduce such instruments in 1994. The various instruments are described in Section 7.5 on policy instruments. Tax incentive support has increased faster than direct funding of R&D during the period 2006-11 and represented over three-quarters of total government funds for R&D in 2011 (Figure 4.14). The extent to which tax incentives benefit the agri-food sector is discussed in Section 7.9.

Figure 4.14. Change in government support for business R&D through direct funding and tax incentives, 2006-11

As a percentage of total support, and annualised growth rates

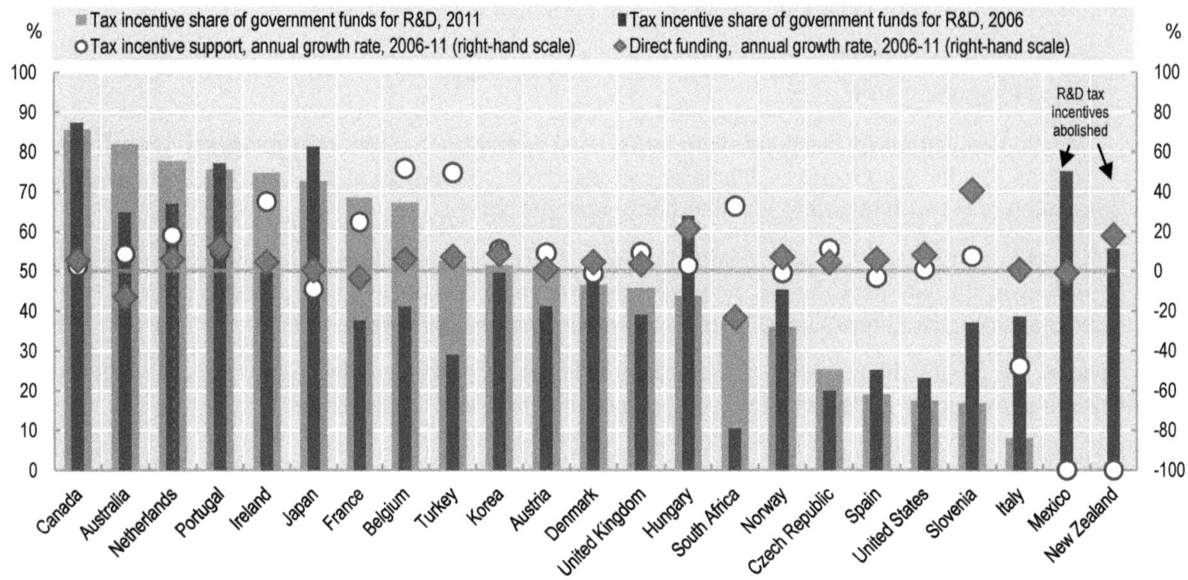

Source: OECD, based on OECD R&D tax incentives questionnaires, January 2010, June 2011 and June 2013, publicly available sources, and OECD, Main Science and Technology Indicators Database, www.oecd.org/sti/msti.htm, June 2013.

StatLink http://dx.doi.org/10.1787/888933251014

According to OECD (2014d), "*The Dutch tax incentives appear to meet most of the principles of good policy design. In particular, there are different brackets, there is a ceiling, there are only small differences in their generosity to profitable and non-profitable firms, and WBSO is regularly monitored and evaluated. The principal question about tax incentives in the Netherlands is not about their design but about the extent of reliance on them and their fitness for purpose, given the diverse challenges involved in raising not only the intensity but also the ambition of firms innovation activity.*"

Summary

- Regulatory barriers to entrepreneurship are the least restrictive amongst OECD countries, and businesses operate in a competitive environment conducive to investment. Regulations remain complex and costly compared to some OECD countries, but the government has set targets to reduce administrative burdens and compliance costs for enterprises, and improve transparency and provision of public services.

- Stringent environmental regulations in the Netherlands are among the strictest of OECD countries. This has contributed to significant improvements in the environmental performance of agriculture, but the ability of agriculture to reach environmental targets could become one of the most important challenges for growth of output and productivity in the sector in the future, thus making the role of eco-innovation central in that regard.

- Competition for land use between agriculture and other activities has increased with the decentralisation of land use policies, and the focus of regional policy on promoting economic development, and move away from restricting land use rather than imposing restrictions.

- On-going efforts to make regulations on products and processes smarter aims to simplify existing legislation, minimise administrative compliance costs and reduce the cost of registering new products and processes, and thus facilitate innovation. A challenge is for regulatory procedures to keep up pace with innovation developments.

- The open trade and investment environment facilitates knowledge flows embedded in agri-food trade and Foreign Direct Investment (FDI), but access to credit for foreign companies limits FDI. The Netherlands is both a significant recipient of agri-food FDI, and a net investor abroad.
- The Dutch financial market is well-developed and the banking sector is large compared to the economic size of the country. But access to finance for innovative firms has decreased since the financial crisis has weakened Dutch banks. In addition, venture capital, which is particularly important for innovative firms, is lacking. The cooperative Rabobank, which is the main supplier of credit to farmers, has provided stable levels of credit to the primary sector, with diverging trends by commodity sub-sector.
- There are multiple programmes to support investment, some being targeted to different stages of innovation. The mix of programmes has changed frequently, and some companies may have difficulty accessing information. To address this issue, information and advice to entrepreneurs is granted for free.
- Tax rates on personal and corporate income are close to the OECD average or median. Few tax rebates specifically target agriculture, although agriculture benefits from tax rebates on green subsidies or investments, and tax exemptions on land transfers or capital gains on land. Glasshouse horticulture benefits from a lower energy tax.
- Tax incentives for innovation account for over three-quarters of government support to business innovation. They benefit mainly sectors with high profit margins and companies with employees involved in Research and Development (R&D).

Notes

1. For more information, see the ondernemersplein's web site at: http://www.ondernemersplein.nl/.
2. The initial objective of crop protection policy as regards ecological risks was to reduce "by 2010, the environmental pressure on surface waters caused by the use of pesticides and herbicides in agriculture and horticulture by 95% compared to that of 1998."
3. Environmental regulation in the Netherlands takes place in the framework of the Environmental Management Act (*Wet Milieubeheer*) and the Decree of general rules governing the environmental management of sites (*Activiteitenbesluit*).
4. OECD Workshop on "Livestock disease policies: Building bridges between science and economics", 3-4 June 2013, Paris. http://www.oecd.org/tad/agricultural-policies/livestock-diseases-2013.htm.
5. The government has recently tabled a plan to address and manage food safety risks in the meat supply chain, following the recommendations of the Dutch Safety Board (OVV) http://www.government.nl/news/2014/06/10/food-safety-paramount-as-government-intervenes-firmly-in-meat-supply-chain.html.
6. Venture capital is a form of private equity. Returns on venture capital investment stem from a trade sale (sale to, or merger with, another company) or an initial public offering in which the company becomes authorized to sell its stock to the general public on a stock exchange. Venture capital funds will not only provide money but will mentor their investee firms (IO, 2012).
7. An angel investor is usually an experienced entrepreneur who provides backing to very early-stage businesses or business concepts.
8. For more information on the Rabobank, see: http://en.wikipedia.org/wiki/Rabobank, https://www.rabobank.com/en/locate-us/europe/netherlands.html.

References

Bergevoet, R. and M. van Asseldonk (2013), "Economic Evaluation of FMD Control Strategies in the Netherlands", paper presented at the OECD Workshop on "Livestock disease policies: Building bridges between science and economics" held on 3-4 June 2013 in Paris. http://www.oecd.org/tad/agricultural-policies/livestock-diseases-2013.htm.

Botta, E. and T. Koźluk (2014), "Measuring Environmental Policy Stringency in OECD Countries: A Composite Index Approach", *OECD Economics Department Working Papers*, No. 1177, OECD Publishing, Paris. DOI: http://dx.doi.org/10.1787/5jxrjnc45gvg-en.

Gerlitz, N. (2012), "The European Union system for health and consumer protection", *Improving Agricultural Knowledge and Innovation Systems: OECD Conference Proceedings*, OECD Publishing, Paris. DOI: http://dx.doi.org/10.1787/.

LEI (2015), LEI Website 'Monitoring mestmarkt aanbod en afzet in Nederland', available at http://www.monitoringmestmarkt.nl/.

Milieu-en Natuurplanbureau (2015), *Tussenevaluatie van de nota Duurzame gewasbescherming*, Report.

OECD (2014a), "Analysing Policies to improve agricultural productivity growth, sustainably: Revised framework". http://www.oecd.org/tad/agricultural-policies/Analysing-policies-improve-agricultural-productivity-growth-sustainably-december-2014.pdf.

OECD (2014b), *OECD Economic Surveys:* Netherlands 2014, OECD Publishing, Paris. DOI: http://dx.doi.org/10.1787/eco_surveys-nld-2014-en.

OECD (2014c), *Water Governance in the Netherlands: Fit for the Future?*, OECD Studies on Water, OECD Publishing, Paris. DOI: http://dx.doi.org/10.1787/9789264102637-en.

OECD (2014d), *OECD Reviews of Innovation Policy: Netherlands 2014*, OECD Publishing, Paris. DOI: http://dx.doi.org/10.1787/9789264213159-en.

OECD (2010), *The OECD Innovation Strategy: Getting a Head Start on Tomorrow*, OECD Publishing, Paris. DOI: http://dx.doi.org/10.1787/9789264083479-en.

OECD (2008), *OECD Rural Policy Reviews: Netherlands*, OECD Publishing, Paris. www.sourceoecd.org/9789264041967.

OECD (2005), *Taxation and Social Security in Agriculture*, OECD Publishing, Paris. DOI: http://dx.doi.org/10.1787/9789264013650-en.

PBL (2012), *Evaluatie van de nota Duurzame gewasbescherming*, Publicatie 500158001, Bilthoven, The Hague.

PBL (2013), *The effects of climate change in the Netherlands*, Amsterdam. http://www.pbl.nl/sites/default/files/cms/publicaties/PBL_2013_The%20effects%20of%20climate%20change%20in%20the%20Netherlands_957.pdf.

Rijksoverheid (2015), Rijksoverheid, website page on Natural Capital, available at: http://www.rijksoverheid.nl/documenten-en-publicaties/kamerstukken/2013/06/22/kamerbrief-over-uitvoeringsagenda-natuurlijk-kapitaal.html.

Valletta, M. (2010), "Consumer Perception and GMOs in the European Union", *Policy Responses to Societal Concerns in Food and Agriculture: Proceedings of an OECD Workshop*, OECD. Available at: www.oecd.org/agriculture/agriculturalpoliciesandsupport/monitoringfarmsupportandevaluatingpolicy/46832852.pdf.

Van Eerdt, M. M., Spruijt, J., van der Wal, E., van Zeijts, H., & Tiktak, A. (2014), "Costs and effectiveness of on-farm measures to reduce aquatic risks from pesticides in the Netherlands", *Pest Management Science*, Vol. 70, N°12, pp. 1840–9. doi:10.1002/ps.3729

World Economic Forum (2013), *The Global Competitiveness Report 2013-2014: Full data Edition*, Geneva 2013. http://reports.weforum.org/the-global-competitiveness-report-2013-2014/#=.

Wossink A. (2003), "The Dutch Nutrient Quota System: Past Experience and Lessons for the Future", paper prepared for the OECD workshop on The Ex-Post Evaluation of Tradeable Permit Regimes.

Chapter 5

Capacity building and public services in the Netherlands

Capacity building, including provision of essential public services, is one of the main channels or incentive areas to support innovation and sustainable development. This chapter concerns three relevant policy areas: infrastructure and rural development policy; labour market policy; and education and skills policy.

The statistical data for Israel are supplied by and under the responsibility of the relevant Israeli authorities. The use of such data by the OECD is without prejudice to the status of the Golan Heights, East Jerusalem and Israeli settlements in the West Bank under the terms of international law.

Infrastructure and rural development policy

Investments in physical and knowledge infrastructure, from Information and Communications Technology (ICT) to transportation facilities, are important for overall growth and development. They are vital to the delivery of and access to important services and play a critical role in linking farmers and related businesses to markets, reducing food waste, boosting agriculture productivity, raising profits, and encouraging investment in innovative techniques and products. Productive and profitable enterprises have higher incentives to invest in sustainable practises that yield long term benefits.

Broader rural development measures also affect agricultural development and structural adjustment. Increased off-farm income and employment opportunities mitigate farm household income risks, facilitate farm investment, and enable a wider range of farm production choices. Improved rural services, from banking to ICT, are important to ensure needed connectivity to suppliers, customers, and collaborators. Rural policy can also attract innovative upstream and downstream industries, with possible spill-over effects locally. By reducing inequalities in economic development and access to services across regions, rural development policies improve the diffusion of innovation (OECD, 2014a).

Physical infrastructure

According to the World Economic Forum's Global Competitiveness Index, the quality of transport infrastructure in the Netherlands is high, with the country ranking 2nd in the OECD area and 6th in the world. The quality of port infrastructure is high, and the country figures in the OECD top 5 for the quality of its roads and air transport infrastructures, but not for the quality of its railroad infrastructure (Figure 5.1.A). In addition, the Netherlands figures in the OECD top 5 for the quality of electricity infrastructure, but not for the quality of telephony infrastructure, due to a lower score for the number of mobile phone subscriptions and fixed telephones (Figure 5.1.B). In terms of internet usage, the Netherlands also figures among the OECD top 5 performers with a high percentage of individuals using internet and a high number of fixed broadband internet subscriptions per capita.

Overview of rural areas

The economic and social differences between regions are modest in the Netherlands. Rural areas in the Netherlands are highly integrated or in close proximity with urban areas (OECD, 2008, 2014b). However, they do not depend on urban areas and offer many opportunities to shop, recreate and make use of public services, which are at a level comparable with urban areas. While the agricultural sector of the Netherlands is performant, rural economies are diversified and do not depend on agricultural activities. Strict and centralised land use planning has facilitated co-existence between agriculture and high density population, but competing demands on land use for dwelling, transport infrastructure and recreation and biodiversity intensify.

Policies have stressed the need to concentrate urbanisation in urban areas, to stimulate inner-city development and to limit urban sprawl in line with the efforts to preserve landscapes and nature reserves. The central government of the Netherlands intends to bring spatial planning decision-making closer to the relevant stakeholders (individuals and private companies), delegating more power to local and provincial authorities (through decentralisation as the first option), and an increasing focus on users (OECD, 2014b).

Figure 5.1. Global Competitiveness Index: Quality of infrastructure, 2013-14

Scale 1 to 7 (best)

A. Transport infrastructure, by component

B. Electricity and telephony, by component

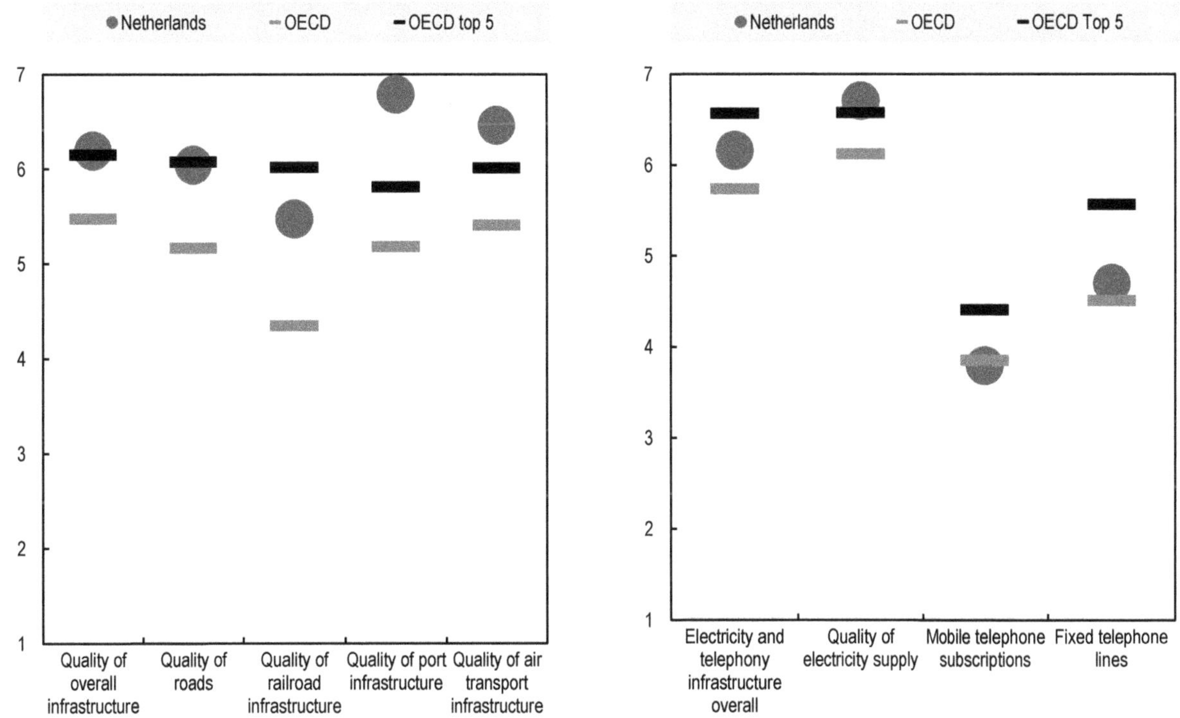

Top 5 refers to the average of the scores for the top 5 performers among OECD countries (France, Germany, Spain, Netherlands and Japan).

Top 5 refers to the average of the scores for the top 5 performers among OECD countries (Luxembourg, Switzerland, United Kingdom, Austria and Germany).

Source: World Economic Forum (2013), *The Global Competitiveness Report 2013-2014*, http://reports.weforum.org/the-global-competitiveness-report-2013-2014/#.

StatLink http://dx.doi.org/10.1787/888933251028

Regional policy

There is currently no national framework for regional policy in the Netherlands. The government recently abandoned the explicit regional policy of the Peaks in the Delta and replaced it by the Enterprise Policy which includes generic policy as well as a policy specifically for innovative sectors, in which the Netherlands excels globally – the Top Sector Policy described in Box 7.1 – rather than focussing on the area-specific strengths of regions. This change was partly driven by the need to prioritise resources in a tight fiscal environment brought by the aftermath of the 2008 global financial crisis and current difficulties in the recovery phase. The main feature of this policy is refocusing on a few key sectors rather than spreading resources over the whole territory. The resources for regional development programmes are therefore expected to come from other institutions, such as the provinces and municipal governments, EU programmes, and other forms of co-operation that would involve partnerships with the private sector (OECD, 2014b).

Strategy for infrastructure and spatial planning

Since June 2012, the National Policy Strategy for Infrastructure and Spatial Planning (*Structuurvisie Infrastructuur en Ruimte* – SVIR) is in force. The SVIR plan represents a strategic agenda for spatial planning policies. One of the aims is laying down the baseline programme of investments. The SVIR sets out a list of national priorities to be followed by the central government. Currently, the national priority identifies three main goals to enhance Dutch competitiveness:

- Ensuring that spatial planning and infrastructure investment support the development of the main clusters (main-ports, brain-port, green-ports, and other urban regions), as well as the development of the country as a whole.
- Improving accessibility.
- Safeguarding the quality of the environment.

The implementation of the central government policy focuses on areas of national interest, for which the central government will take responsibility. Among these areas of national interest is a robust main road, rail, and waterway network around and between the most important urban regions, including connections with the hinterland.

Responsibility for balancing urban and green areas at regional level is left to the provincial authorities. To this end, the central government has abolished the national landscape policy and reduced the number of nature management regimes. The government also strengthened the cohesion between the various modes of transport and between spatial development and mobility.

One of the instruments in relation to the SVIR is the **Multi-Year Plan for Infrastructure, Spatial Planning and Transport (MIRT)**, an investment programme set up by the national government. The MIRT aims to improve the coherence between investments in spatial planning, the economy, mobility and liveability at the national level. It is a national programme which contributes to the regional agenda, providing long-term investment framework for the Netherlands and the regions.

In spatial terms, the Dutch economy consists of several **"ports" and "valleys"**. In these structures there is a strong component of co-operation among public institutions, knowledge institutions, and businesses. Main ports, green ports and valleys are especially relevant for the agri food complex (Box 5.1). The designation of Greenports (by the National Spatial Strategy 2004) has led in particular to the development of a strategic vision on the future of the horticulture sector (*Vitaal Tuinbouwcluster 2040 – een Toekomststrategie voor Greenport Holland*) addressing issues like knowledge and innovation, room for development, accessibility, regulation and instruments, the European agenda, and the restructuring of the sector.

Box 5.1. Main ports, green ports and valleys relevant for the agri-food complex

Main ports

The sea port of Rotterdam (the "front door" to the European market), Seaports Amsterdam (the fourth port in Europe and the second in terms of value added) and Schiphol airport in the province of Noord-Holland, near Amsterdam. This area is also known as the Randstad region. The ports are connected to the rest of the country, and the rest of Europe, through an infrastructure network which exploits both terrestrial and fluvial routes.

Green ports

These are based on the agro-food sector and contribute to an important share of total Dutch exports. Several green ports have been developed:

- Greenport Westland – Oostland (Zuid-Holland): This area is the largest international greenhouse horticulture area in the Netherlands.
- Greenport Venlo: This green port will concentrate on markets like Germany and Central Europe.
- Greenport Aalsmeer: On a limited acreage here floriculture, breeding, propagation, production, sale and trade are concentrated. A logistical advantage is achieved by the presence of Schiphol Airport.
- Greenport Duin en Bollenstreek: This greenport focuses more on the cultivation of flower bulbs. The Greenport has a great attraction for tourists through the bulb fields and the presence of the Keukenhof. It is one of the Greenport areas where several functions, such as housing, employment and tourism are linked.
- Greenport Boskoop: This Greenport is well known for tree and shrub crops.
- Greenport Noord-Holland Noord, also called Agriboard.

Box 5.1. Main ports, green ports and valleys relevant for the agri-food complex (*cont.*)

In addition to the core clusters, there are a number of large-scale production (satellite) areas which together form Greenport Holland.

Valleys

The food valley in the province Gelderland consists of a network of food companies, research institutes and universities. The aim is to create conditions for food manufacturers and knowledge institutes to work together and join their efforts in developing new and innovating food concepts.

The seed valley Enkhuizen aims to combine knowledge and business in maintaining the position as world leader in seed products and propagation materials. Thirty-five technology driven firms can be found in a small area around the provincial town of Enkhuizen. Local based R&D is of central importance for the vitality of these firms.

Rural development

The former objectives of the rural development policy were laid down in the *Agenda voor een Vitaal Platteland* (Agenda for a Living Countryside) which was adopted in 2004. The Agenda specified the policy tasks for the economic, ecological and socio-cultural aspects of rural life. The national government concluded administrative agreements with the individual provinces for the performance of these tasks in the period between 2007 and 2013. The financial resources of the ministries and other parties, including the provinces, the European Union, municipalities, water boards, social organisations and private individuals, were aggregated in the *Investeringsbudget Landelijk Gebied* (ILG, the Rural Area Investment Budget). Government funding of approximately EUR 3.5 billion was made available for the 2007-13 period. The provinces had the responsibility for the implementation of the ILG.

However, pursuant to the financial measures implemented by the Rutte I Government (October 2010 – April 2012) the government has terminated the ILG administrative agreements. Settlement agreements were concluded by the government with all provinces at the end of 2012. The provinces have been allocated a budget for nature management, the implementation of the decentralised nature policy and the fulfilment of the mandatory legal obligations arising from the ILG period.

Rural Development Programmes

The Rural Development Programmes in the Netherlands are established within the framework of EU regulations, in particular the EU Common Agricultural Policy (CAP). Of the EUR 973 million of public funding (EU and national) available under the Rural Development Plan 2007-13, 30% was dedicated to Axis 3 to encourage diversification in agriculture, strengthen micro-enterprises in rural areas and improve access to the countryside and promote rural tourism, with another 10% for rural development projects relying on a multi-sectoral approach and local partnerships to address specific local problems (LEADER approach).

As part of the newly introduced Partnership Agreement for 2014-20, EU member states are required to indicate how they will strengthen co-ordination and complementarity between the different programmes. (Compared to the previous programming period, a higher share of CAP Pillar 2 funds (EAFRD) is dedicated to environment-related measures (Table 5.1). See Chapter 6.4 for more information on rural development programmes in the CAP and their implementation in the Netherlands.

Table 5.1. Allocation breakdown by thematic objective and by fund (except territorial co-operation), 2014-20

Million EUR

	ERDF	ESF	EAFRD	EMFF	Total
01. Strengthening research, technological development and innovation	332. 5		22. 1		354. 6
02. Enhancing access to, and use and quality of, information and communication technologies					.
03. Enhancing the competitiveness of small and medium-sized enterprises, the agricultural sector (for the EAFRD) and the fisheries and aquaculture sector (for the EMFF)			162. 3	21. 9	184. 2
04. Supporting the shift towards a low-carbon economy in all sectors	121. 3				121. 3
05. Promoting climate change adaptation, risk prevention and management			176. 2		176. 2
06. Preserving and protecting the environment and promoting resource efficiency			176. 2	73. 6	249. 7
07. Promoting sustainable transport and removing bottlenecks in key network infrastructures					.
08. Promoting sustainable and quality employment and supporting labour mobility	11. 7	126.			137. 7
09. Promoting social inclusion, combating poverty and any discrimination	21. 6	361.	41.		423. 5
10. Investing in education, training and vocational training for skills and lifelong learning			7. 4		7. 4
11. Enhancing institutional capacity of public authorities and stakeholders and an efficient public administration					.
Technical assistance	20. 3	20. 3	22. 2	6. 1	68. 8
Total	**507. 3**	**507. 3**	**607. 3**	**101. 5**	**1723. 5**

EAFRD: European Agricultural Fund for Rural Development ; EMFF : European Maritime and Fisheries Fund.

Source: EU Commission web site : http://ec.europa.eu/contracts_grants/pa/partnership-agreement-nederlands-summary_en.pdf.

StatLink http://dx.doi.org/10.1787/888933251362

Labour market policy

Labour market policy influences employment composition and labour mobility, in particular by facilitating (or discouraging) labour to adapt to new circumstances. It can play an important role in facilitating structural adjustment, including farm consolidation, by assisting excess labour in farming to exploit more remunerative non-farm income and employment opportunities. Policy on international mobility of human resources can also help to better match labour supply with demand, and can affect innovation and knowledge transfer through exchange of skills and skilled labour (OECD, 2014a).

Main performance indicators

After access to finance, restrictive labour regulations are cited as the second most important barrier for doing business in the Netherlands. According to 2013 OECD indicators, employment protection legislation in the Netherlands is amongst the most restrictive of OECD countries (OECD, 2013). According to WEF Global Competitiveness Index, which is based on an executive opinion survey, labour market efficiency in the Netherlands is above the OECD average in general, but below for wage flexibility determination, for hiring and firing practices, and for pay and productivity. On the other hand, it ranks among the OECD top 5 for reliance on professional management and country capacity to retain talent, two aspects that are important to foster innovation (Figure 5.2).

Figure 5.2. Global Competitiveness Index: Labour market efficiency, 2013-14

Scale 1 to 7 (best)

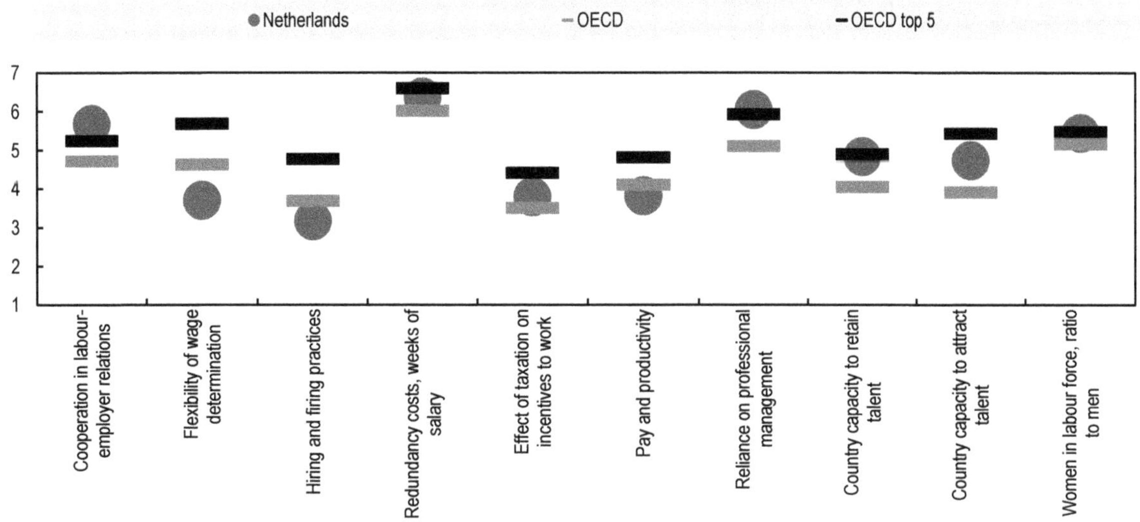

OECD top 5 refers to the average of the scores for the top 5 performers among OECD countries (Switzerland, United States, United Kingdom, Canada and New Zealand).

Source: World Economic Forum (2013), *The Global Competitiveness Report 2013-2014*, http://reports.weforum.org/the-global-competitiveness-report-2013-2014/#.

StatLink http://dx.doi.org/10.1787/888933251037

Over the years, some of the inertia in labour markets has been reduced, and efforts to increase flexibility continue. OECD surveys have emphasised that plans to lower and cap severance payments and simplify dismissals would improve the labour market and would contribute to SME dynamism (OECD, 2012, 2014c).

The authorities will reduce the protection of permanent contracts and simultaneously increase the protection of employees on temporary contracts. In July 2014 a new law on labour and security (*Wet Werk en Zekerheid*) was approved by the parliament. The main changes (from 1 July 2015) are as follows:

- Unemployment: the maximum duration of unemployment insurance paid by the Government will gradually be reduced from 38 months to 24 months.
- Dismissal: As of 1 July 2015 there will be two dismissal procedures, via the Employee Insurances Implementing Agency (UWV) in case of dismissals for economic reasons and long-term disability. All other cases are settled by the sub-district court.
- Flex: Employees with temporary contracts are entitled to receive a permanent contract after two years, instead of three years.
- Transition payment/compensation: All employees with at least two years of service at the time of (not voluntary) dismissal are entitled to a transition payment which can be used for retraining or outplacement in order to find another job or profession.

Availability of labour force

For employers in the agriculture and food industry the availability of labour force is a serious concern. Kupper et al. (2012) state that, according to the Council for Agricultural Vocational Education, there is a market potential of 15% for green education while the actual enrolment is only about 5%. "Improving the branches" image and being a good employer' is one of the three central themes in the Human Capital Agenda (HCA, 2012). Various efforts are being made to improve the availability of labour force, such as positive campaigns targeted at young people. Also, employers in agri-food and farmers collaborate with schools developing life-long-learning programmes, providing internships and explaining career possibilities to students.

Many employers in agri-food, especially those in intensive livestock farming, mushroom production and greenhouse vegetable production have great concerns about the low profitability of their operations in previous years (Berkhout et al., 2014). An increasing number of these business owners have to do their utmost to avoid bankruptcy. As labour costs represent an important share (around 20%) of the total costs in these branches (Berkhout et al., 2014), business owners are looking for ways to reduce these costs. However, the poor image of the various branches in agriculture and horticulture does not make it easy to attract motivated employees at low cost, all the more because the employment rate in the Netherlands is relatively high (it exceeds 75% compared to an OECD average of 65%) (OECD, 2014c). As a result employers in agri-food hire (seasonal) migrant workers from various EU member states, especially from Eastern Europe, to do the job.

Figure 5.3 (Ooijevaar et al., 2013) shows the increase, including the seasonal fluctuations, in migrant workers in the Netherlands from 2007 till 2010. To avoid unfair competition with the Dutch labour market, the Dutch government requests a level playing field and verifies that migrant workers are paid the same wages as Dutch workers. These conditions restrict low-skill immigrants, but not so much highly-skilled workers, Yet, modifications in labour market regulations are needed recurrently in order to deal with creative arrangements to reduce labour costs, such as using workers employed by a foreign employment agency under the legal provisions of the country of origin or providing (mandatory) housing facilities to migrant workers at (too) high rents (PvdA, 2014).

Figure 5.3. Number of immigrants from new Eastern-Europe EU member states.

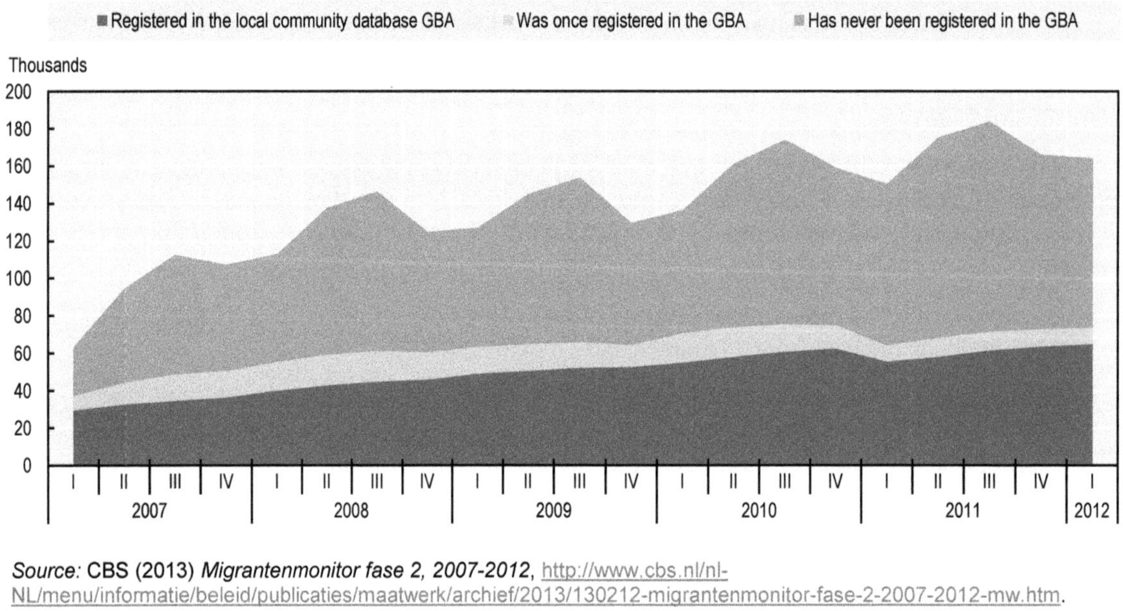

Source: CBS (2013) *Migrantenmonitor fase 2, 2007-2012*, http://www.cbs.nl/nl-NL/menu/informatie/beleid/publicaties/maatwerk/archief/2013/130212-migrantenmonitor-fase-2-2007-2012-mw.htm.

StatLink http://dx.doi.org/10.1787/888933251040

The agri-food sector also requires skilled workers. Efforts to meet the demand for highly-skilled workers consists in improving the skills in the working population through education and training, and attracting foreign workers with matching skills in the Netherlands (see next section).

Education and skills policy

Education policy affects innovation in at least three ways: a high level of general and scientific education facilitates acceptance of technological innovation by society at large; innovation systems require well-educated researchers, teachers, extension officers, and producers to develop relevant innovations; it is generally easier for farmers and business operators with higher education and skills to adopt some technological innovations (OECD, 2014a).

To compensate for the relatively high costs of labour and land in the Netherlands, and to keep a "license to produce", the agricultural and food complex has to operate on a high level of performance with respect to productivity and innovation, while at the same having to deal with societal demands in a responsible manner. This is a challenge because complying to stringent environmental demands, for example, is much more complicated and costly in densely-populated areas than it is in spacious areas. Therefore, a well-functioning agricultural knowledge and innovation system is essential, but this can only exist if it is formed by people with a high level of general and scientific education and skills.

Overall achievement

The Dutch population benefits from a high quality education system, which performs well in international comparison (Nutsche et al., 2014), and achieves good results in terms of educational attainment, and innovation skills.

According to WEF Global Competitiveness Index, which is based on an executive opinion survey, the Netherlands is among the top 5 OECD countries in terms of the overall performance of higher education and training (Figure 5.4). It ranks particularly high in terms of the quality of higher education, which reflects business executives' assessment of: how well the educational system meets the needs of a competitive economy; the quality of math and science education in schools; the quality of business schools; and the prevalence of widespread Internet access in schools. According to business executives, provision of on-the-job training, i.e. the availability of high-quality, specialised training services and the extent to which companies invest in training and employee development, is also among the top 5 of OECD countries. The quantity of education in terms of enrolment rates is comparable to the OECD average, which scores high in any case.

OECD figures on the share of the population that have attained tertiary education is also close to the OECD average (Figure 5.A1.1). These figures also show that a higher percentage of the younger age group (25-34 year-olds) has attained tertiary education. This may be the result of the additional investments in primary, secondary and post-secondary education but there may be alternative explanations for continuing education, such as changing attitudes towards education, the economic recession with high unemployment or changing job requirements.

Tertiary education attainments yield better employment prospects in the Netherlands than in other OECD countries. The employment rate of workers with tertiary education is 22 percentage points higher than the employment rate of workers with upper secondary education. Nevertheless, the employment prospects of workers with a secondary education are relatively high when compared to other OECD countries.

According to the OECD Programme for International Student Assessment (PISA) (OECD, 2014d), the Netherlands scores are relatively high on the reading, math and science skills of 15-year-old students (Figure 5.5), despite the relatively low total average expenditure on education as a percentage of GDP. Similarly, according to the OECD survey of adult skills (PIAAC),[1] adults in the Netherlands show above-average proficiency in literacy, numeracy and problem solving in technology-rich environments compared with adults in the other countries participating in the survey. In particular, about three-quarters of the adult population in the Netherlands displays moderate to good skills and readiness to use information and communication technologies (ICT) for problem solving, the highest proportion among surveyed countries, together with Norway and Sweden, as well as one of the highest frequencies in using ICT at home (OECD, 2014e).[2] According to the same survey, the Netherlands appears to have little mismatch between the literacy and numeracy proficiency of workers and the demands of their jobs.

The Dutch population is also highly skilled in languages. This facilitates cross-country co-operation and social innovation. According to an EU survey of Europeans and their languages, 94% of respondents from the Netherlands said they were able to speak at least one other language besides their mother tongue (compared to 54% at the EU level). English was the most commonly spoken of these foreign languages, but 71% also responded that their German was good enough to hold a conversation. The Netherlands is also one of the eight member states of the EU in which a majority of respondents said that they had practical skills in at least two foreign languages (European Commission, 2012).

Figure 5.4. Global Competitiveness Index: Higher Education and Training, 2013-14

Scale 1 to 7 (best)

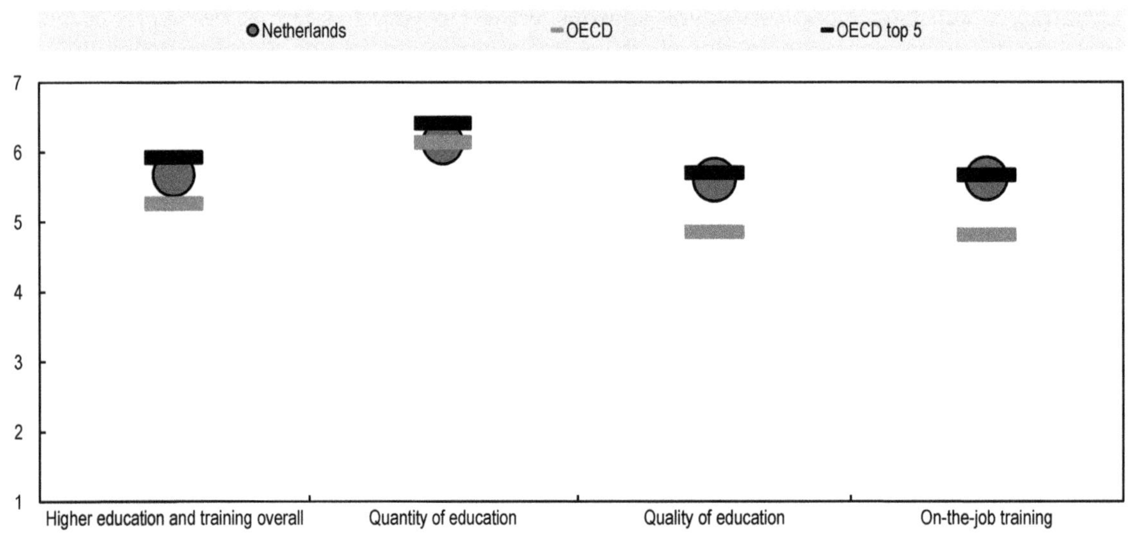

Top 5 refers to the average of the scores for the top 5 performers among OECD countries (Finland, Germany, Switzerland, Belgium and Netherlands).

The quantity of education index is based on secondary and tertiary education enrolment rates from UNESCO Institute for Statistics. The quality of education index is based on responses from a WEF Executive Opinion Survey on "How well does the educational system meet the needs of a competitive economy; Executives' assessment of the quality of math and science education in schools and the quality of business schools; and on how widespread is Internet access in schools. The on-the-job-training index is based on survey responses on the availability of high-quality, specialised training services and the extent to which companies invest in training and employee development.

Source: World Economic Forum (2013), *The Global Competitiveness Report 2013-2014: Full data Edition*, Geneva 2013. Executive Opinion Survey; Data for the Quantity of education index comes from UNESCO Institute for Statistics. http://reports.weforum.org/the-global-competitiveness-report-2013-2014/#=.

StatLink http://dx.doi.org/10.1787/888933251056

Figure 5.5. OECD PISA 2009 mean scores by the Netherlands, United States and OECD average

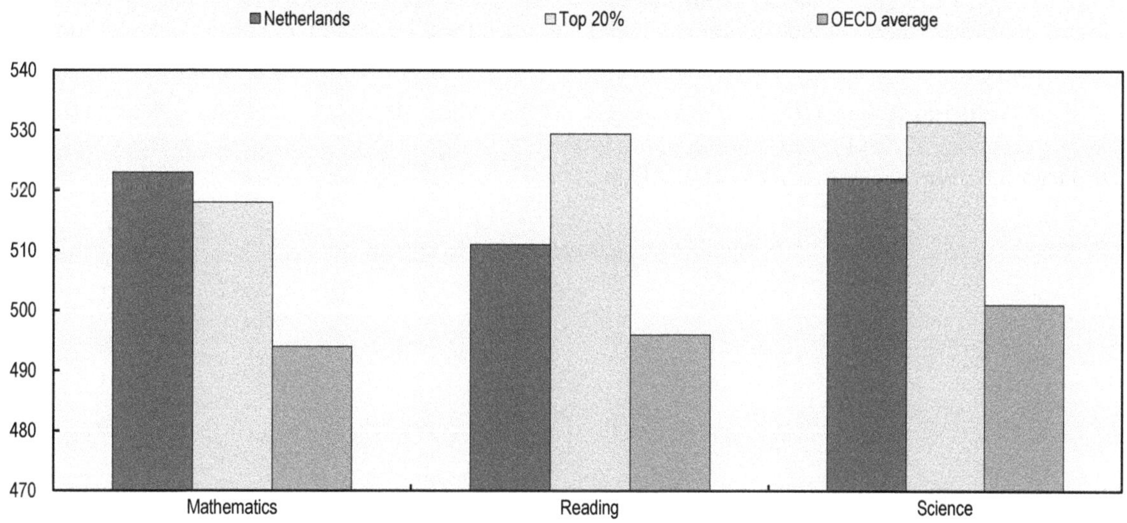

Source: OECD (2014d), *PISA 2012 Results in Focus*, http://www.oecd.org/pisa/keyfindings/pisa-2012-results-overview.pdf.

StatLink http://dx.doi.org/10.1787/888933251066

Trends in education expenditures

There is no straightforward correlation between education expenditure and outcome: while students and adult skills are well-above the OECD average, expenditure on education as a percentage of GDP is close to the OECD average. Total expenditures on education per student are higher than the OECD average (Figure 5.A1.4). Total expenditures as a percentage of GDP in the Netherlands have increased faster than the OECD average, as they were below in 2000 and slightly above in 2011 (Figure 5.A1.3). Despite this faster growth, they are still much lower than in neighbouring countries such as Belgium, the United Kingdom and Denmark.

Expenditures per student increased less than expenditure as a percentage of GDP over the period 2000-12 (Figure 5.A1.5). In fact, the increase in expenditure as a percentage of GDP is partly due to the decline in GDP. The change in expenditure on primary, secondary and post-secondary education in the Netherlands can be explained by a higher expenditure per student, whereas the increase in expenditure on tertiary education is mainly due to the higher number of students (OECD, 2013).

Current policy efforts

According to the most recent OECD *Education Policy Outlook* (OECD, 2014f), "current policy efforts focus on maintaining quality in tertiary education and responding to emerging labour market needs. The top sectors' human capital agendas encourage co-ordination to identify and prepare for emerging skill needs. In 2013, the government launched the 2020 National Technology Pact, involving major stakeholders. Co-operation between higher education institutions, vocational secondary education and the business sector is a main aspect of the Pact, which aims to increase the number of technically trained people."

The Dutch Green Education system

In the Netherlands, agricultural education is embedded in the so called **green education** (agriculture, nature and food) and is organised through close co-operation with the agricultural sector, under the responsibility of the Ministry of Economic Affairs, while general education is under the responsibility of the Ministry of Education.

The green education is subject to the regulatory framework of the Dutch education system (Figure 5.6).[3] Thus, the secondary education includes pre-vocational secondary education (VMBO) programmes (four years), which combine general and vocational education and prepare pupils for senior (also referred to as upper) secondary vocational education and training (MBO – four years).

The Netherlands has 110 green schools (76 VMBO/MBO establishments and 33 comprehensive schools that offer VMBO-green training), coordinated in 12 Agricultural Education Centres (AOCs) and one Regional training centre (ROC) with MBO-green.

Figure 5.6. The Dutch education system

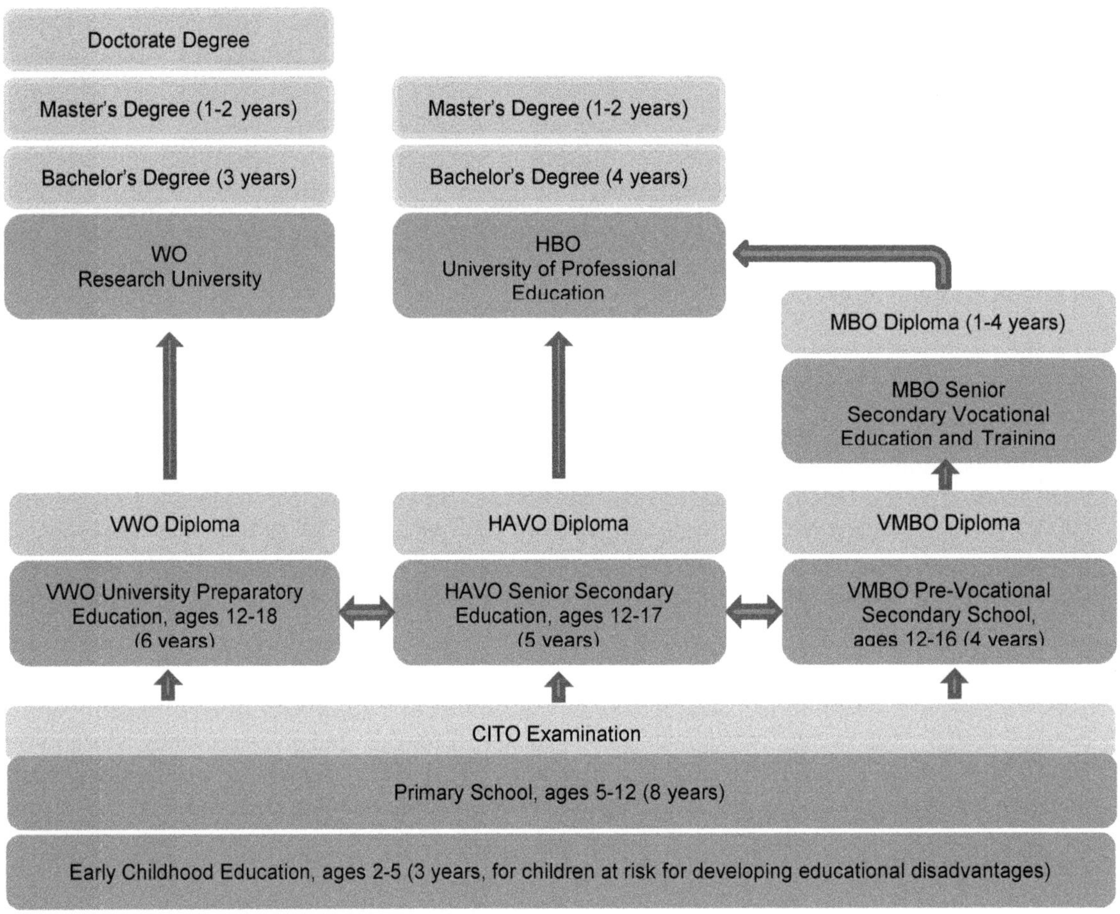

CITO: Central Institute for Test Development; HAVO: Senior general secondary education; HBO: Higher Professional Education; MBO: Secondary vocational education; VMBO: Pre-vocational secondary education; VWO: Pre-university secondary education; and WO: University education. Acronyms checked in http://www.oecd.org/edu/school/NLD_CBR_Evaluation_and_Assessment.pdf.

Source: Caggiano (2014), "AKIS and advisory services in The Netherlands. Report for the AKIS inventory (WP3) of the PRO AKIS project". Online resource: www.proakis.eu/publicationsandevents/pu.bs .

Higher education is provided by two types of institutions: Research universities (WO) and Universities of applied sciences (HBO). Research universities are primarily focused on research-oriented programmes, while HBO are more practice oriented, offering programmes of higher professional education to prepare students for specific professions. There are 5 HBOs providing green curricula (4 HBOs-green and one university of applied sciences with a green department) and only one green Research University (Wageningen University). Utrecht University has a Faculty of Veterinary Medicine.[4]

In addition, the Practical Training Centre (PTC+) provides supplementary and specialist education for horticulture, livestock and other specialised areas. This private organisation, which is part of the Aeres Group, offers courses and training programmes not only in the Netherlands but all over the world. IPC Groene Ruimte offers similar services with focus on the themes: Trees, Fauna, Green, Soil and Water.

Enrolment in green education

According to the Dutch Inspectorate of Education, in the year 2011/12 there were 78 300 green education students: 31 700 enrolled in VMBO, 30 500 in MBO, 9 100 in HBO and 7 000 in Wageningen University (Caggiano, 2014).

Kupper et al. (2012) have compared the enrolment in green and non-green education, based on figures of the year 2008/09 (Table 5.2).

Table 5.2. Non-green and green students enrolment, 2008/09

	VMBO	MBO				HBO	WO
		Level 1	Level 2	Level 3	Level 4		
Non-Green	212 229	21 600	134 100	136 600	216 500	383 883	213 900
Green	31 488	2 142	6 479	7 114	11 399	7 909	6 149
Ratio Green/ non-green (%)	14. 8	9. 9	4. 8	5. 2	5. 3	2. 1	2. 9

HBO: Higher Professional Education; MBO: Secondary vocational education; VMBO: Pre-vocational secondary education; WO: University education.

Source: Kupper, H.; Laurentzen, R. & Mulder, Martin (2012), "Recent Policy Developments in Green Education in the Netherlands", *The Journal of Agricultural Education and Extension*, DOI: http://dx.doi.org/10.1080/1389224X.2012.655966.

StatLink http://dx.doi.org/10.1787/888933251379

They conclude that there are relatively few students that choose green education (a total of 72 677 students). According to the Council for Agricultural Vocational Education, there is a market potential of 15% of all students enrolled, whereas actual enrolment is about 5%. Enrolment at all educational levels increased in 2008/09 after a slight decrease in the previous years. Since their previous 2006 article, enrolments at green academic institutes rose by 25% (compared to 12% for all universities), at green higher professional level by 3% (10% at all higher professional level), green vocational level by 15% (5% at all vocational level) and the prevocational level dropped by 9% (all pre-vocational decreased by 7%).

As the ratios of green to non-green student enrolment in Table 5.2 indicate, the attractiveness declines from low- to high level green education. Compared to the vocational level, the ratios for higher professional and academic are low. While the level of education of the general working population increased between 2001 and 2009 this might not be the case in some green sectors where the majority of the vacancies are still open to employees with no basic qualifications: 32% in the husbandry sector, 48% in the agricultural sector, 38% in the horticultural sector and 47% in the food sector (Kupper et al., 2012). The composition of enrolment in green education reflects market demand, which is higher at VMBO level.

Recent developments

Nowadays, many students who graduate at a vocational education level (MBO) continue their education at a higher professional level. This has probably to do with the economic recession and the low availability of jobs at the MBO level. There is also a sharp increase in national and foreign student figures at Wageningen University and the schools for higher professional green education (Box 7.3). The reason for this may be found in the increasing (media) awareness for food, feed and fuel issues ("feeding 9 billion people in 2050") and sustainability issues at the work level, which leads to good economic and job perspectives in the green sector. For example, 80% of students find jobs quickly, and remain in the sector. An alternative explanation may be that green and non-green education increasingly become interrelated, for example in the fields of food technology, biotechnology and biobased economy. The green HBOs and Wageningen University then become an option for students who have less connection with agriculture. Moreover, the green HBOs and Wageningen University are relatively small and have a personal approach. This is an aspect that many candidate students (and their parents) appear to appreciate.

Comparing the 2011/12 enrolment figures of the recent AKIS report (Caggiano, 2014) with the 2008/09 figures of the Kupper et al. (2012) article we see that VMBO enrolment in green education has increased by 0.6%, MBO by 12%, HBO by 15% and Wageningen University by 14%. Most recent data from the Education Executive Agency (DUO) indicate that the growth of Wageningen University has even accelerated. The education budget for Wageningen University comes from the Ministry of Economic Affairs and is limited to a growth of 2% each year. This means that the existing increase in student numbers has been met by reducing expenditure per student. It also means that the growth in student numbers cannot be indefinite unless more money is channelled to green education, including for extending the capacity of the University in terms of physical infrastructure and staff.

Methods to deliver education are also changing: life-long learning and distance learning programmes are developing rapidly, allowing for a larger potential student base. The need to focus vocational education on entrepreneurship, training students to become managers rather than employees, is also being increasingly recognised.

The Green Table

Various actors in the Green education system are looking for ways to improve the system, despite the economic situation in the Netherlands with less government expenditures. Kupper et al. (2012) mention the foundation of the Green Knowledge Co-operative as an organisational structure, providing the sector with adequate and up-to-date knowledge and offering attractive learning environments. The current situation, however, is that the Green Knowledge Co-operative was dismantled in 2014 and other institutions were created.

To maintain a connection between the various educational institutions, as formerly institutionalised within the Green Knowledge Co-operative, "the Green Table" was founded in 2014. This roundtable with representatives from the educational institutions aims to collaborate on three main themes, guarding common interests: 1) in discussion and negotiation with the government; 2) in the relationship education – labour market; and 3) for maintaining a good knowledge infrastructure (Boetzkes, 2014).

Human capital agenda in top sectors

The **Top Sector policy**, which defines the R&D strategy, includes the development of a Human Capital Agenda (Box 7.1). In the two top sectors dedicated to food, agriculture and horticulture, stakeholders from agribusiness, education, research and government together developed this Agenda (HCA, 2012). The central idea behind the Human Capital Agenda is to get more involvement and responsibility of agribusiness in education and skills development and, in attracting sufficient students at various levels, to ensure an adequate supply of qualified employees in agriculture and horticulture.

Three important themes have been identified: 1) improving the branches' image and being a good employer, 2) developing a job-oriented curriculum, and 3) promoting life-long-learning. To stimulate the involvement of agribusiness, the stakeholders have created Centres for innovative craftsmanship (at the MBO level) and Centres of expertise (at the HBO level). The business plans of these centres reveal that these centres will be co-financed (in kind and in cash) by the agribusiness partners gradually in the next four to five years. In return tools, trainings and internships will be provided, and students and teachers plan to execute (research) projects for the agribusiness partners (CoE Agrodier, 2013; CIV Agri & Food, 2013).

These agendas, thus, offer further opportunities to adapt skills to industry needs, and for the education institutions to obtain funding from the private sector, although these are very limited in the secondary institutions as private industries are more interested in research and make contact with high-skilled (potential) students. Moreover, the budget available for co-funding activities of Centres is limited.

Summary

- Economic activities and rural populations benefit from an excellent infrastructure network and good access to public services.
- There is a strong demand for unskilled seasonal workers by agriculture and horticulture, which has been increasingly met by temporary migrants. There is also strong demand in the food processing and horticultural sector for skilled workers and for upgrading skills, which is being addressed in collaboration with the education system.
- Co-ordination between agri-food research, education and industry has been long standing and is being strengthened. But delays in the response to growing industry needs may lead to temporary shortages of skills.
- The Dutch education system is highly-ranked at the world level and it is responsive to business needs. The agri-food, as other sectors, benefits from a well-educated labour force. This contributes to facilitating innovation in firms and acceptance of innovation in the society.
- The education system offers strong and attractive degrees in agriculture, food and nature management, concentrated in specialised institutions. As a result, the number of students enrolled in agricultural technical and higher education increases faster than the budget allocated to "green" education. The concentration of agriculture-related activities into specialised institutions facilitates multidisciplinary approaches with agricultural specialists, but it may limit exchanges with specialists from other domains, which are increasingly important for agriculture to address current and future challenges.

Notes

1. Survey of adult skills – Country Note: Netherlands: www.oecd.org/site/piaac/Country%20note%20-%20Netherlands.pdf.
2. OECD. Survey of Adult Skills (PIAAC). (www.oecd.org/edu/eag.htm). http://dx.doi.org/10.1787/888933115065, Chart A1.a., Chart A1.b.
3. Centre on International Education Benchmarking.
4. Utrecht University: Faculty of Veterinary Medicine: www.uu.nl/en/organisation/faculty-of-veterinary-medicine.

References

Berkhout, P., H. Silvis and I. Terluin (2014), *Landbouw-Economisch Bericht 2014*, LEI-rapport 2014-013, LEI Wageningen UR, The Hague, June, 200 pp. http://edepot.wur.nl/306953.

Boetzkes (2014), "Bestuurders tekenen voor Groene Tafel", Vakblad Groen Onderwijs, 56(2), 7 March 2014, pp. 30-31. http://edepot.wur.nl/298046.

Caggiano, M. (2014), "AKIS and advisory services in The Netherlands. Report for the AKIS inventory (WP3) of the PRO AKIS project." Online resource: www.proakis.eu/publicationsandevents/pubs.

CBS (2013), *Migrantenmonitor fase 2, 2007-2012*. www.cbs.nl/nl-NL/menu/informatie/beleid/publicaties/maatwerk/archief/2013/130212-migrantenmonitor-fase-2-2007-2012-mw.htm.

CIV Agri & Food (2013), "Centrum Innovatief Vakmanschap Agri & Food" Businessplan 2013–2017, juni 2013. Online resource: www.civ-agri-food.nl/over-civ-agro-food/businessplan.aspx.

CoE Agrodier (2013), "Meer Waarden met Dieren". Business Plan Centre of Expertise Agrodier, 23 mei 2013. Online resource: http://edepot.wur.nl/282363.

European Commission (2012), *Europeans and their languages*, Special Eurobarometer 386, June. http://ec.europa.eu/public_opinion/archives/ebs/ebs_386_en.pdf.

HCA (2012), "Human Capital Agenda Agri & Food en Tuinbouw & Uitgangsmaterialen". Online resource: http://hetgroeneplein.nl/file/view/11094142/hca-human-capital-agenda-agri-and-food-en-tuinbouw-and-uitgangsmaterialen.

Kupper, H, R. Laurentzen, and M. Mulder (2012), "Recent Policy Developments in Green Education in the Netherlands", *The Journal of Agricultural Education and Extension*, Vol. 18, No. 2, pp. 121-139, DOI: http://dx.doi.org/10.1080/1389224X.2012.655966.

NCEE (2014), Center on International Education Benchmarking: The Netherlands Overview. www.ncee.org/programs-affiliates/center-on-international-education-benchmarking/top-performing-countries/netherlands-overview/.

Nusche, D., et al. (2014), *OECD Reviews of Evaluation and Assessment in Education: Netherlands 2014, OECD Reviews of Evaluation and Assessment in Education*, OECD Publishing, Paris. DOI: http://dx.doi.org/10.1787/9789264211940-en.

OECD (2014a), "Analysing Policies to improve agricultural productivity growth, sustainably: Revised framework". www.oecd.org/tad/agricultural-policies/Analysing-policies-improve-agricultural-productivity-growth-sustainably-december-2014.pdf.

OECD (2014b), *OECD Territorial Reviews: Netherlands 2014*, OECD Publishing, Paris. DOI: http://dx.doi.org/10.1787/9789264209527-en.

OECD (2014c), *OECD Economic Surveys: Netherlands* 2014, OECD Publishing, Paris. DOI: http://dx.doi.org/10.1787/eco_surveys-nld-2014-en.

OECD (2014d), *PISA 2012 Results in Focus*: www.oecd.org/pisa/keyfindings/pisa-2012-results-overview.pdf.

OECD (2014e), *Survey of adult skills – Country Note: Netherlands*: www.oecd.org/site/piaac/Country%20note%20-%20Netherlands.pdf.

OECD (2014f), *OECD Science, Technology and Industry Outlook 2014*, OECD Publishing, Paris. DOI: http://dx.doi.org/10.1787/sti_outlook-2014-en.

OECD (2013), *Education at a Glance 2013: OECD Indicators*, OECD Publishing, Paris. DOI: http://dx.doi.org/10.1787/eag-2013-en

OECD (2012), *OECD Economic Surveys: Netherlands* 2012, OECD Publishing, Paris. DOI: http://dx.doi.org/10.1787/eco_surveys-nld-2012-en.

OECD (2010), *The OECD Innovation Strategy: Getting a Head Start on Tomorrow*, OECD Publishing, Paris. DOI: http://dx.doi.org/10.1787/9789264083479-en.

Ooijevaar, J.; Sluiter, N., and Verschuren, S. (2013), "Immigranten en werknemers uit de Europese Unie in Nederland". CBS Bevolkingstrends 2013, December 2013, 14 pp. www.cbs.nl/nl-NL/menu/themas/bevolking/publicaties/bevolkingstrends/archief/2013/2013-immigranten-en-werknemers-uit-de-europese-unie-in-nederland-pub.htm.

PvdA (2014), "Actieplan 'Fatsoenlijke huisvesting arbeidsmigranten' (action plan 'decent housing migrant workers')", 13 March. www.pvda.nl/berichten/2014/03/Fatsoenlijke+huisvesting+arbeidsmigranten.

World Economic Forum (2013), *The Global Competitiveness Report 2013-2014: Full data Edition*, Geneva 2013. http://reports.weforum.org/the-global-competitiveness-report-2013-2014/#=.

Annex source

OECD (2014), *Education at a Glance 2014: Highlights*, OECD Publishing, Paris. DOI: http://dx.doi.org/10.1787/eag_highlights-2014-en.

For further information on Green education

Blekemolen, M. and J. Groenendijk (2013), 'Éindrapport Beleidsdoorlichting Groen Onderwijs' (Final Report on Policy Review of Green education, Ministry of Economic Affairs, December. www.rijksoverheid.nl/documenten-en-publicaties/rapporten/2014/06/02/beleidsdoorlichting-groen-onderwijs-eindrapport.html.

Annex 5.A1

Expenditure on education

Figure 5.A1.1. Population that has attained tertiary education, 2000, 2012

Percentage of 25-64-year olds who have been through tertiary education

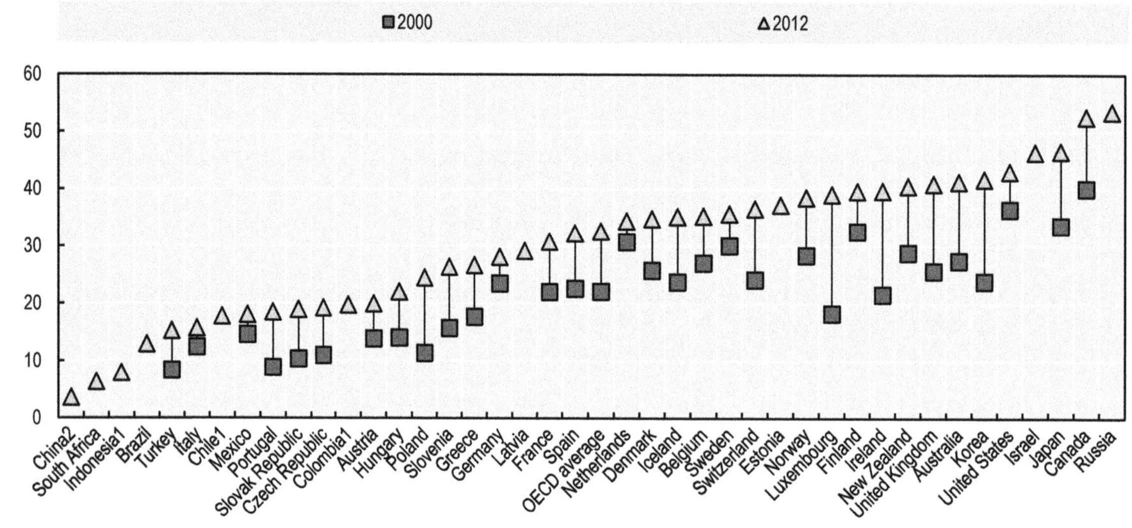

Source: OECD (2014), *Education at a Glance 2014*, Chart A1.1, available at http://dx.doi.org/10.1787/888933114951.

StatLink http://dx.doi.org/10.1787/888933251073

Figure 5.A1.2. Public and private expenditure on educational institutions as a percentage of GDP, 2011

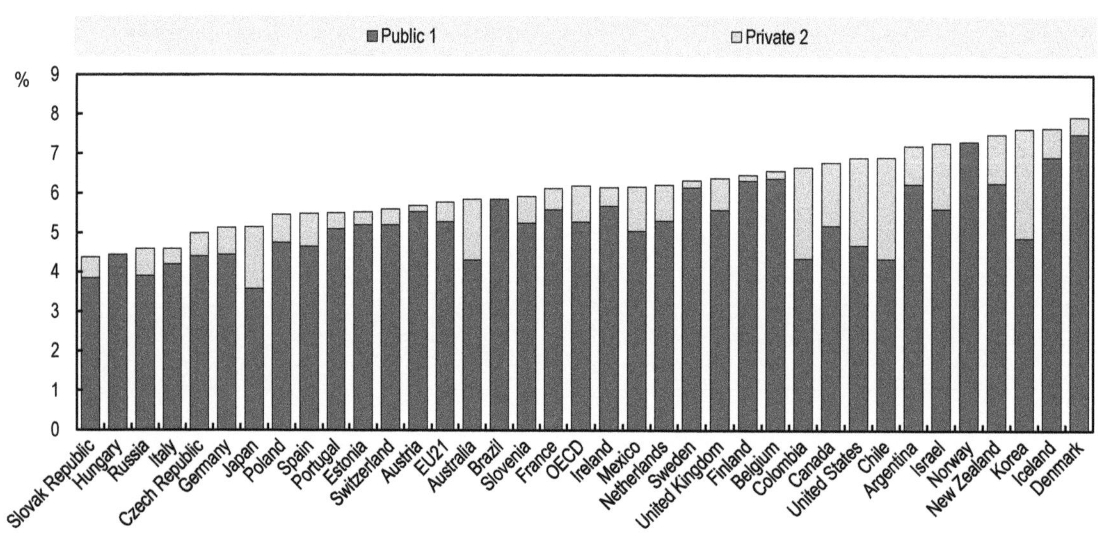

Data for Canada refer to 2010 and for Chile 2012. EU21: EU member states that are members of the OECD.

1. Public expenditure includes public subsidies to households attributable for educational institutions, and direct expenditure on educational institutions from international sources.

2. Private expenditure is net of public subsidies attributable for educations institutions.

Source: OECD (2014), *Education at a Glance 2014: OECD Indicators*, OECD Publishing, Paris. DOI: http://dx.doi.org/10.1787/eag-2014-en.

StatLink http://dx.doi.org/10.1787/888933251082

Figure 5.A1.3. Expenditure on educational institutions as a percentage of GDP for all levels of education, 2000, 2008 and 2011

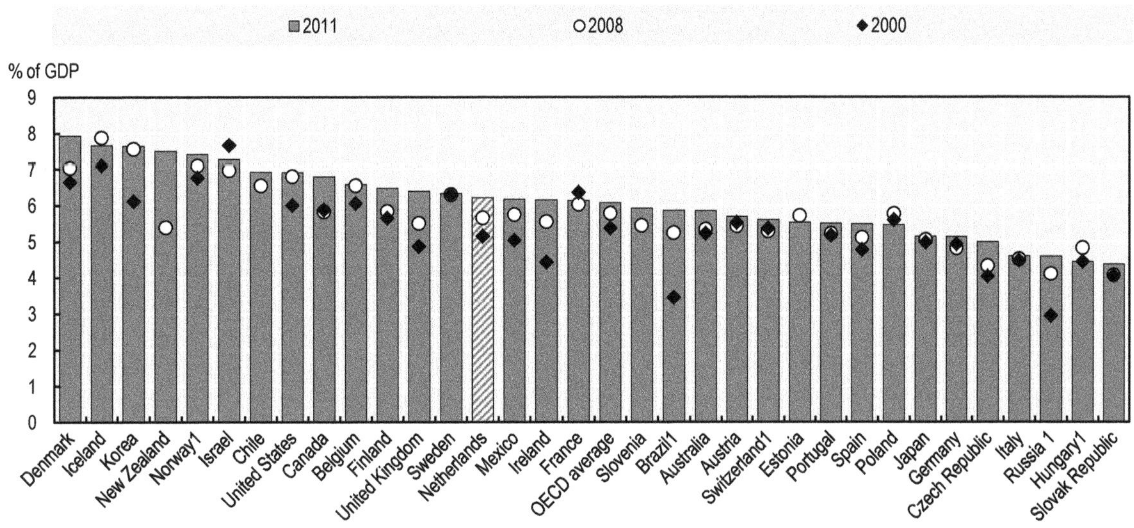

Source: OECD (2014), *Education at a Glance 2014*, Chart B2.1, available at http://dx.doi.org/10.1787/888933117288.

StatLink http://dx.doi.org/10.1787/888933251099

Figure 5.A1.4. Average public expenditure per student by educational institutions, 2011

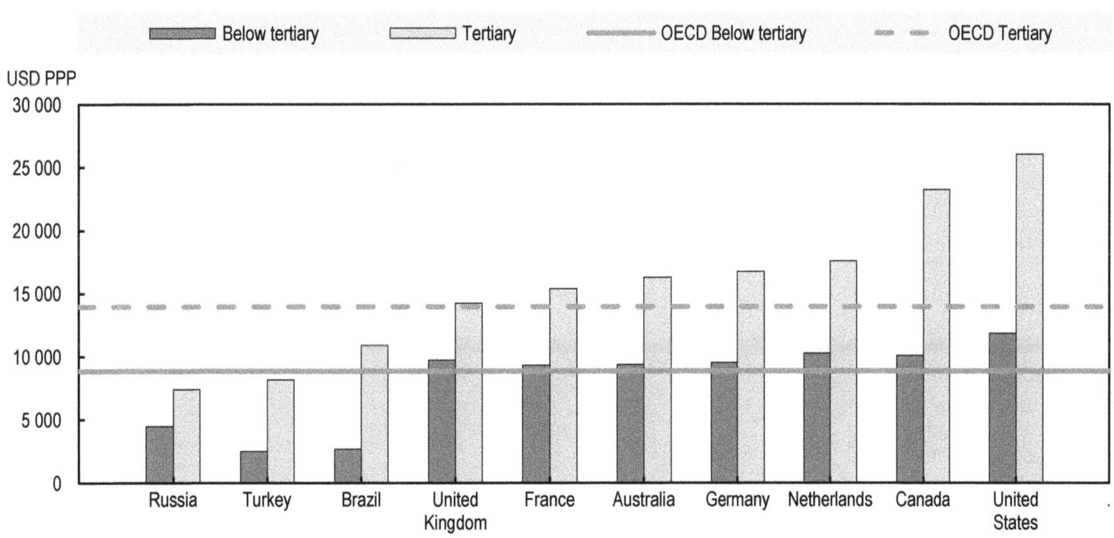

Source: OECD (2014), *Education at a Glance 2014: OECD Indicators*, OECD Publishing, Paris.
DOI: http://dx.doi.org/10.1787/eag-2014-en.

StatLink http://dx.doi.org/10.1787/888933251103

Figure 5.A1.5. Change in Expenditure per student 2000, 2008 and 2011

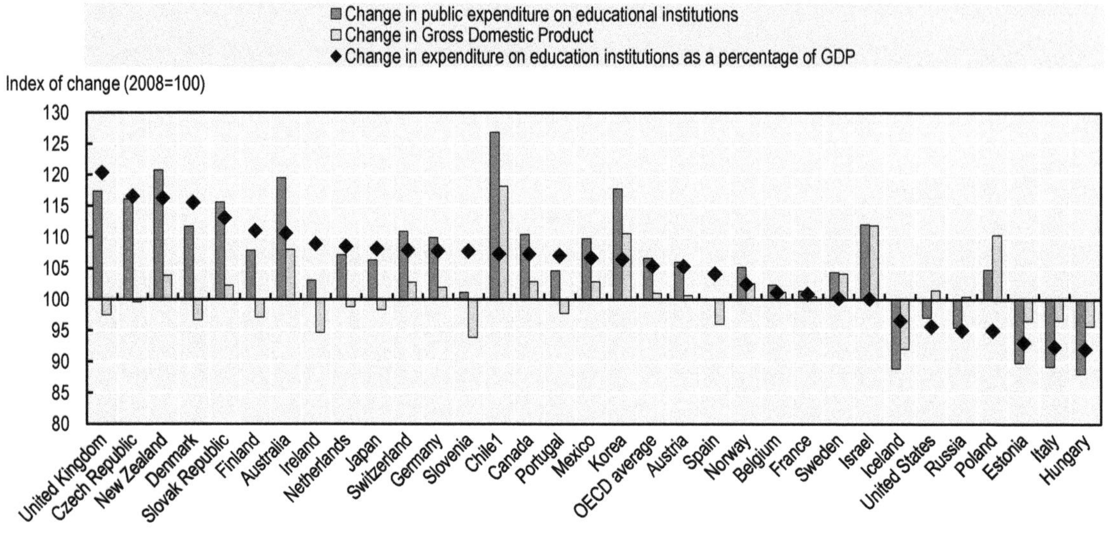

Source: OECD (2014), *Education at a Glance 2014*, Chart B2.1, available at http://dx.doi.org/10.1787/888933117326.

StatLink http://dx.doi.org/10.1787/888933251112

Chapter 6

Dutch domestic and trade-related agricultural policy

This chapter provides an overview of the framework for agricultural policies, which is the Common Agricultural Policy (CAP), the main instruments and options at the EU level, the recent CAP reform for 2014-20, and its implementation in the Netherlands. It also refers to the provision of general services supported by the government. It finally discusses policy impact on innovation for productivity and sustainability.

The statistical data for Israel are supplied by and under the responsibility of the relevant Israeli authorities. The use of such data by the OECD is without prejudice to the status of the Golan Heights, East Jerusalem and Israeli settlements in the West Bank under the terms of international law.

Dutch policies with regard to agriculture, food and rural areas are largely the result of decisions taken at the European Union level. As a member state, the Netherlands shares the responsibility for the development and implementation of the common policies. The goals of the national policy fit within the context of European policies. There are several European policies that concern the agri-food complex. Non agriculture-specific policies are relevant to agriculture, food and rural areas, such as environmental policy (Nitrates Directive, 1991), research and development policy ("framework programmes"), industry policy and transport policy. Elements of further European integration included increased attention to competition policy. More recently, the European Union has reformed its food policy (General Food Law) (Oskam et al., 2011) and is currently working on the so-called Plant and Animal Health Package.

The description of the measures below draws on information from earlier OECD publications (OECD, 2011, 2014), updated using information from the EU Agricultural and Rural Development web site.[1]

Fundamentals of the CAP

Agriculture has been part of the European integration process since its integration into the Common Market. The Common Agricultural Policy (CAP), launched in 1962, initially focused on internal market integration and protecting EU agriculture from world markets.[2] Later on, issues of external integration (world trade relations and policies) became more prominent. Successive reforms initiated since 1992 have reduced support and changed the way it is delivered to farmers. Border protection has decreased with the implementation of the Uruguay Round Agreement on Agriculture (URAA) from 1995 to 2001. Direct intervention on domestic markets was gradually reduced and replaced by direct payments to producers, which became less and less linked to specific commodities and to current production (OECD, 2011).

The present CAP is composed of two pillars. Pillar 1 entails a Common Market Organisation (CMO), which lays out rules for providing market price support measures, as well as broad-based direct payments. Pillar 1 measures are fully funded by the European Union, through the European Agricultural Guarantee Fund (EAGF) funds. Pillar 2, or Rural Development Regulation (RDR) of Agenda 2000, includes a list of available measures from which member states can choose. EU funding for Pillar 2 measures comes from the European Agricultural Fund for Rural Development (EAFRD) and requires co-funding at the national or regional levels. EU co-financing rates vary by country and by type of measure (higher for agri-environmental measures, lower for investment support in EU15). National (or Regional) Rural Development Plans (RDPs) are the basis for the implementation of RDR measures. They specify the list of measures chosen and the associated EU funding and national/regional co-funding for a period of seven years. Plans established by member states need to be approved by the Commission.

Over the period 2007-13, pillar 1 accounted for about 80% of all CAP EU expenditures. Since the end of the 2000s, direct payments account for most of Pillar 1 expenditures, as successive commodity market reforms and high world price levels reduced expenditures to support domestic markets.

In 2013 an agreement on CAP reform for 2014-20 was reached. While continuing on the path of reform started in the early 1990s, for the first time the entire CAP was reviewed all at once and the European Parliament acted as co-legislator with the Council (Section 6.5).

The change in the orientation of the CAP is demonstrated by the evolution of expenditure, echoing the policy shift since 1992, away from product based support towards producer support and considerations for the environment (Figure 6.1). In 1992 market management represented over 90% of total CAP expenditure, driven by export refunds and intervention purchases. By the end of 2013 it dropped to just 5% as market intervention has become a safety net tool for times of crisis and direct payments are the major source of support; 94% of which are decoupled from production. From 2014 onwards, the allocation of direct payments dedicated to coupled support, young farmers, small farmers, etc. will depend upon the choices made by member states.

From 2014 onwards, the annual budget available for the CAP stabilises in current terms and decreases in constant terms. On average, Pillar 2 is foreseen to receive a slighted larger share of CAP expenditures. Of a total amount of about EUR 362.8 billion in 2011 prices for 2014-20, over 76.6% is foreseen for Pillar 1 and the remaining 23.4% for Pillar 2. However, the share of expenditure between pillars may change, with the possibility for member states to transfer up to 15% of their national envelopes between pillars, enabling them to better target spending to their specific priorities.

Figure 6.1. Developments in CAP expenditure

Current prices, calendar year

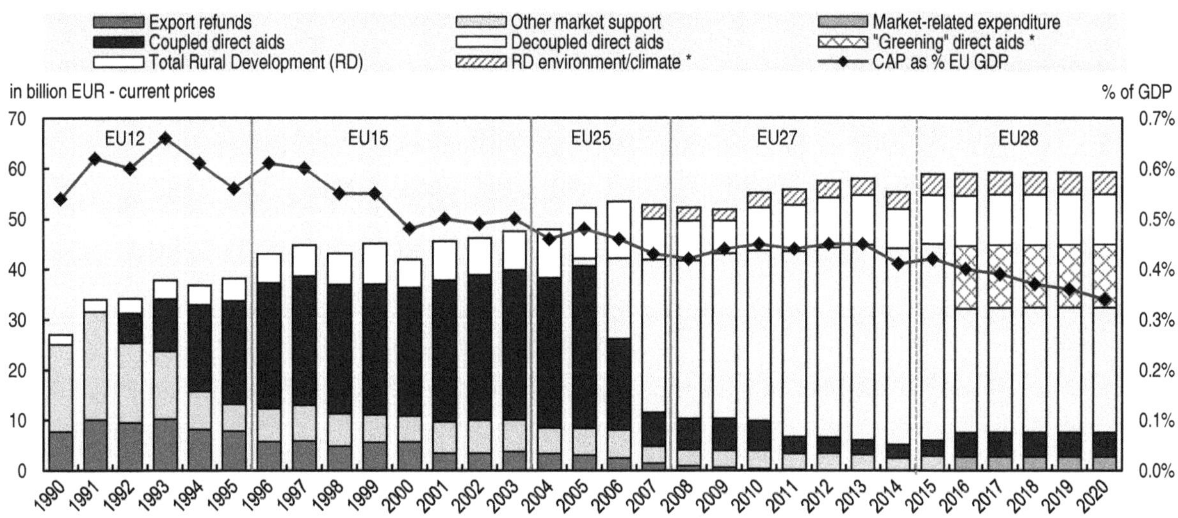

* Precise amounts for 2014-2010 may vary according to member state implementation plans

Source: DG Agriculture and Rural Development, http://ec.europa.eu/agriculture/cap-post-2013/graphs/graph3_en.pdf.

StatLink http://dx.doi.org/10.1787/888933251128

Price and income support measures (Pillar 1)[3]

Direct payments

Direct payments in Pillar 1 are mostly granted in the form of a single payment, which is not linked to current production, land use or other input use. Member states can choose, however, to keep some Pillar 1 payments based on current commodity production parameters (so-called "coupled" payments). Until the implementation of CAP 2014-20 in 2015, in EU15 member states and some new member states, the single payment was delivered through the Single Payment Scheme (SPS), which, depending on the country, was based on individual farmers' historical entitlements (historical implementation model) or (in some cases partly) delivered as a uniform payment per hectare within a region (regional model). For the implementation of the 2003 CAP reform, the Netherlands opted for the historical implementation model and chose to implement a limited set of coupled payment (for calves and cattle, nuts and protein crops until 2010; for potato starch, flax and hemp, flaxseeds and the processing of dried fodder until 2012). Specific payments delivered in the Netherlands under Article 68 during 2010-14 supported the adoption of practices to improve traceability, risk management, sustainable and animal welfare friendly production methods, as well as high quality agriculture.

As Pillar 1 payments result from historical references depending on past production or hectares, current levels whether expressed as a percentage of gross receipts or per hectare differ by member state. As other countries with a high density of livestock production, the Netherland has one of the highest per hectare rates of pillar 1 payments (OECD, 2011, Figure 2.2). On the other hand, payments relative to gross farm receipts are particularly low, as Dutch agriculture is very productive and large parts of agricultural production is not supported by direct payments (e.g. horticulture) (OECD, 2011, Figure 3.3).

The full granting of all EU direct payments is linked to adherence to environmental standards, as well as standards related to food safety, animal and plant health, and animal welfare. In addition, member states must ensure that all agricultural land is "kept in good agricultural and environmental condition". Minimum standards in this respect are drawn up at the national level. This provision called "cross-compliance" applies to all EU payments.

Market price support measures

Market price support measures include border measures such as import tariffs and export subsidies, and domestic measures such as production quotas, administered prices, intervention purchase, and assistance to private storage. The implementation of those measures is governed by a single CMO of agricultural markets established in 2007 to group the previous 21 CMOs for commodity sub-sector. The CMO also includes various aid schemes such as aid for processing or for consumption, and provides rules concerning marketing and production standards, the recognition of producer and operator organisations by member states, and competition within the Common market, including rules applying to state aids.

Table 6.1 provides a snapshot of market price support measures available or applied in 2012/13, within indicating the size of the intervention. For example, expenditures on export subsidies have dramatically declined (OECD, 2011). Similarly, private storage support, which has replaced public intervention for many products, is hardly used. An important development is the end of dairy production quotas in April 2015, as well as the planned removal of sugar production quotas in 2017/18, which will strengthen the market orientation of the sector and foster adjustment to market signals.

Pillar 2 measures for the programming period 2007-13 focused on three "thematic axes" corresponding to policy objectives attributed to measures within each axis: 1) improving the competitiveness of the agricultural and forestry sectors; 2) improving the environment and the countryside; and 3) improving the quality of life in rural areas and encouraging diversification of the rural economy.

Table 6.1 Market price support measures in 2012/13

	Cereals	Oilseeds	Sugar	Fruits and vegetables	Milk/ dairy	Beef	Sheep	Pig	Poultry/ eggs
Import tariffs	X		X	X	X	X	X	X	X
Tariff rate quotas	X		X	X	X	X	X	X	X
Export subsidies	X		X	X	X	X		X	X
Production quotas			X[2]		X[2]				
Public intervention	X[1]				X	Xo			
Private storage			Xo		Xo	Xo	Xo	Xo	
Market withdrawals			Xo	X					
Commodity specific consumer subsidies				X	X				

x: exist in 2012/13. xo not currently used.

1. Ceiling set to 3m tonnes for common wheat at a purchase price of EUR 101.31 per tonne, then by tendering.

2. The milk quota system has been gradually phased out over the period 2009/10 and 2013/14 until it expired in April 2015. The sugar quota system is planned to expire in 2017/18.

Source: WTO notifications; PSE database; OECD, 2014.Rural development measures (Pillar 2).

- Axis 1 includes measures for farm modernisation, the setting-up of young farmers, early retirement, semi-subsistence farms undergoing restructuring, vocational training, producer groups, adding value to farm and forestry products, and restoring production potential damaged by natural disasters.
- Axis 2 includes agri-environmental and animal welfare payments, payments to farmers in areas with handicaps, payments for afforestation, payments for protecting biodiversity in specific sites, and support to non-productive investments.
- Axis 3 groups measures encouraging the diversification into non-agricultural activities, tourism activities, the creation and development of microenterprises, rural services, and the conservation of rural heritage.
- Pillar 2 also supports a fourth axis of projects using the "LEADER approach", relying on a multi-sectoral approach and local partnerships to address specific local problems; as well as technical assistance.

Measures in Axis 1 and 2 are almost exclusively for farmers, while any local actor can apply for measures in Axis 3 and 4. A provision of the programming was that Axis 1 and 3 should account for at least 10% of EAFRD funds, Axis 2 for 25% and, for EU15 member states, the LEADER axis for 5%. Following European Commission (EC) rules, Member states develop Rural Development Programmes (RDPs), indicating the type of measures chosen as well as EU and national or regional funds allocated to each axis. Within EU rules, there is flexibility in the implementation of specific measures.

Axis 1 measures account for about a third of all expenditures, Axis 2 slightly less than 50% and other Axes about 20%. Some measures are selected in all member states: vocational training, farm modernisation, payments to farmers in non-mountainous areas with handicaps, and agri-environmental measures, which are the only mandatory one.

In 2013-14 on average, the Netherlands spent over a third of EU Pillar 2 funds on rural development, non-sectoral, measures in Axes 3 and 4, a share much higher than in other EU15 member states (Figure 6.2). This indicates a stronger focus on non-sectoral measures than in all EU15 member states. Moreover, 10% of Pillar II funds is spent on projects implemented using a LEADER approach, as was indicated in the RDP for 2007-13. On the other hand, the use of Axis 1 and Axis 2 measures is below the EU15 average in the Netherlands.

Figure 6.2. Distribution of EAFRD expenditure by axis, 2013-14

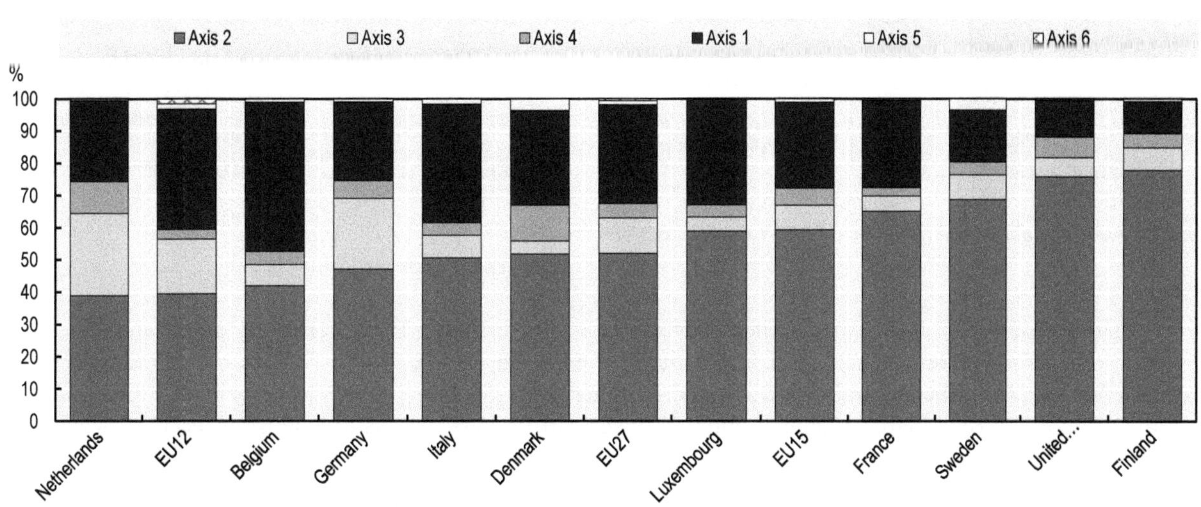

Source: Author's calculations based on EAFRD expenditure (February 2015).

StatLink http://dx.doi.org/10.1787/888933251130

Figure 6.3. Main Pillar II measures by axis, 2013-14

Percentage share of main measures in funds for in each axis, 2013-14 average expenditure

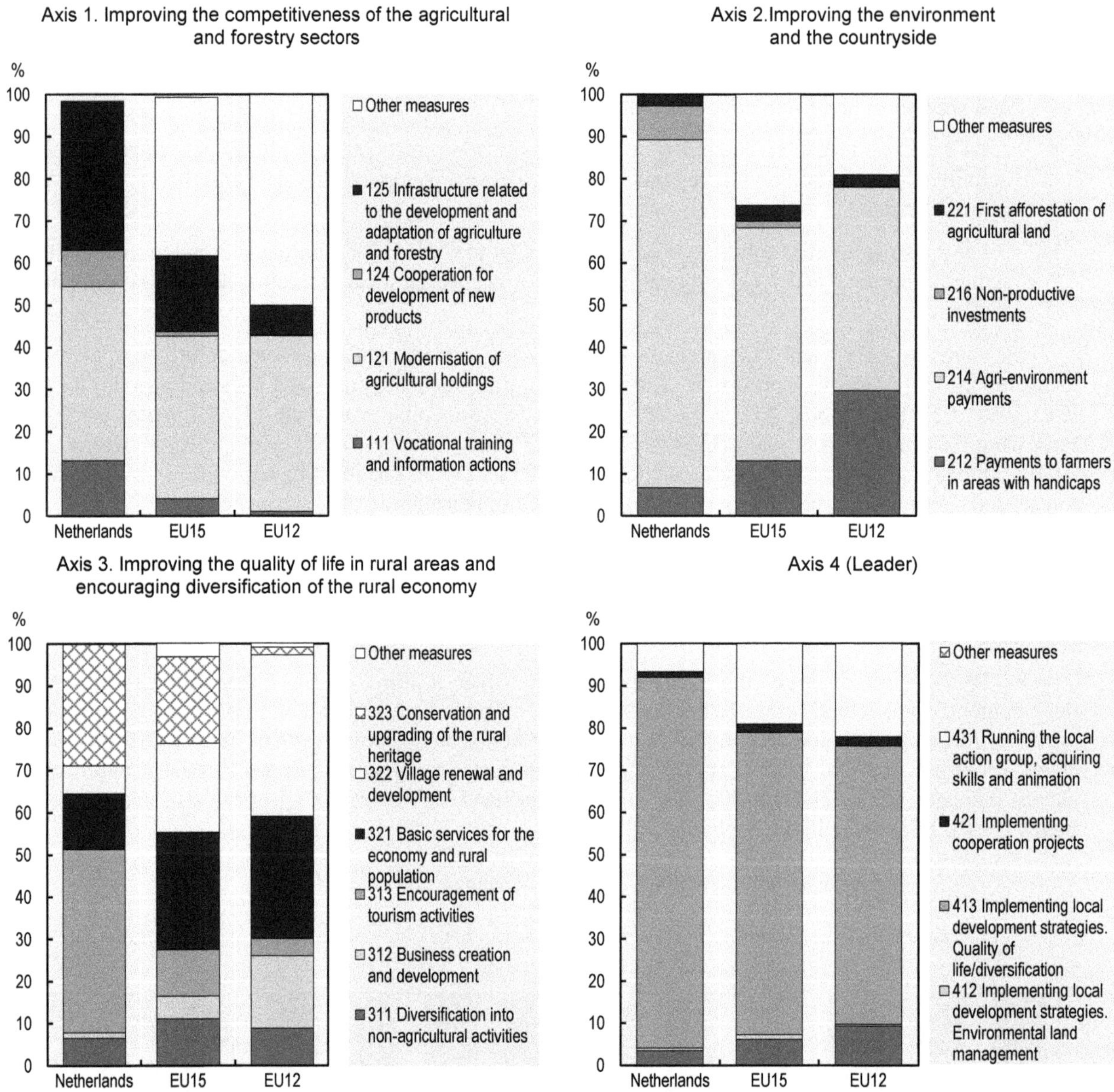

Source: Author's calculations based on EAFRD expenditure (February 2015).

StatLink http://dx.doi.org/10.1787/888933251146

In the 2007-13 RDP, the Netherlands chose to focus funds on a relatively small number of measures, in particular under Axis 1 and Axis 2, which offer a large choice (Figure 6.3). Compared to the EU15 average, the Netherlands dedicated a higher share of Axis 1 funds to vocational training and infrastructure development. A small amount is to support farmers' access to extension services. A large part of Axis 3 funds two measures, one for the conservation of local heritage, and the other one to encourage rural tourism. Most of LEADER funds are to improve quality of life and diversify the local economy. Over 80% of Axis 2 funds were dedicated to agri-environmental measures, compared to 55% on average in the EU15 and 48% in the EU12.

Agri-environmental measures are mainly for nature conservation by agriculture. The primary objective is to increase biodiversity and nature in the National Ecological Network. Adapted farm management can increase the farmland biodiversity. This involves developing and managing species-rich grasslands or strips of grasslands around cultivated fields, on soils that are poor in nutrients. These eco-systems are also an important habitat for many insects, small mammals, etc. The species-rich grasslands occupy either land plots or the edges of fields. In the latter case, they also form a buffer zone to a watercourse, thus hindering the superficial transport of minerals and other substances to the water.

CAP reform 2014-20

The most recent reform process, accompanying the wider reworking of the EU's Common Agricultural Policy (CAP), was essentially completed in December 2013 with the approval of the basic legislative acts for 2014-20. The CAP continues to be composed of two pillars. It will be possible for member states to transfer funds both ways between the two pillars within specific limits and conditions. The overall budget for the CAP is lower than the previous envelope and amounts to EUR 363 billion over six years in 2011 prices.

Direct payments

The CAP for 2014-20 gives some discretion for member states in implementing both pillars. EU member states submitted their plans to the European Commission in 2014. Member states choices for Pillar 1, and rural development plans are expected to be approved by the Commission by mid-2015.

The new CAP contains important elements of redistribution of Pillar 1 direct payments, both within and between member states (OECD, 2014). By introducing so called internal convergence of direct payments under the Basic Payment Scheme there is a progression towards flat rates per hectare at national or regional level. So-called external convergence of payments applies the same principle to progressing towards flatter payment rates between member states. Further fine-tuning of the basic Payment Scheme includes a reduction of amounts paid per recipient above a certain threshold, the option to grant higher payments to the first hectares per recipient, a compulsory top-up of payments to young farmers and simplified procedures for small farmers who receive only small amounts of direct support.

While the direct payment system has become "flatter" under the CAP 2014-20, member states are granted more flexibility to define the specific implementation, in particular of measures under Pillar 2. This flexibility includes the possibility to use an increased share of up to 13% of the national budget envelope for commodity-specific payments and in addition 2% can be allocated to protein crops.

A focus on improving the environmental performance of agriculture is introduced by making 30% of the direct payment entitlement contingent on certain farming practices, including crop diversification, maintenance of permanent pastures and the establishment of ecological focus areas. Alternatively, member states have flexibility to implement national certification schemes instead. In addition, existing cross-compliance conditions are redefined and continue to apply for the direct payments. As the green direct payment is compulsory it has the advantage of introducing practices that are beneficial for the environment and climate on most of the utilised agricultural area. Figure 6.4 illustrate the cumulative roles of cross-compliance, green payments and pillar 2 measures in the greening of EU agriculture.

Figure 6.4. The new greening architecture of the CAP

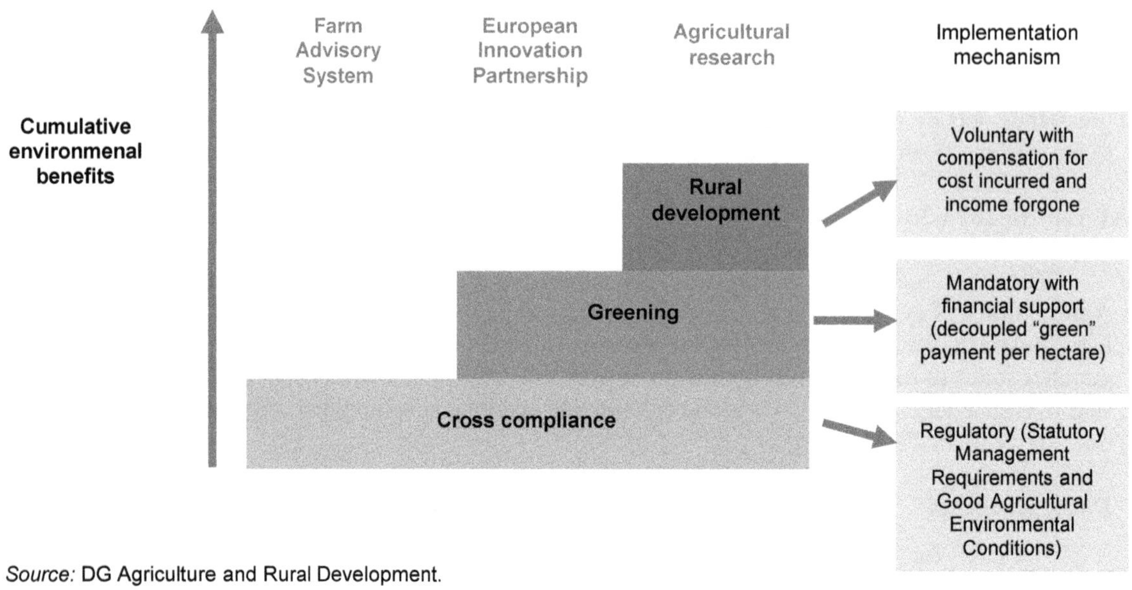

Source: DG Agriculture and Rural Development.

Rural development measures

In line with the overall CAP objectives three long-term strategic objectives for EU rural development policy in the 2014-20 period can be identified: 1) fostering the competitiveness of agriculture; 2) ensuring the sustainable management of natural resources, and climate action; and 3) achieving a balanced territorial development of rural economies and communities including the creation and maintenance of employment. They broadly correspond to the three axis of the previous programming period.

The 2014 reform leaves in place many of the key features of rural development policy from 2007–13. In particular, as in the past, the policy is implemented through national and/or regional rural development programmes (RDPs) which run for seven years. However, overall, the 2014 reform brings change by: improving the strategic approach to constructing RDPs; strengthening the content of rural development measures; simplifying rules and/or reducing the related administrative burden where possible; and linking rural development policy more closely to the other European Structural and Investment (ESI) funds. Member states had to build their RDPs based upon at least four of the six common EU priorities.

- Fostering knowledge transfer and innovation in agriculture, forestry and rural areas.
- Enhancing the viability and competitiveness of all types of agriculture, and promoting innovative farm technologies and sustainable forest management.
- Promoting food chain organisation, animal welfare and risk management in agriculture.
- Restoring, preserving and enhancing ecosystems related to agriculture and forestry.
- Promoting resource efficiency and supporting the shift toward a low-carbon and climate-resilient economy in the agriculture, food and forestry sectors.
- Promoting social inclusion, poverty reduction and economic development in rural areas.

In turn, each rural development priority identifies more detailed areas of intervention ("focus areas"). Within their RDPs, Member States / regions set quantified targets against these focus areas, on the basis of an analysis of the needs of the territory covered by the RDP. They then set out which measures they will use to achieve these targets and how much funding they will allocate to each measure. Funding is drawn partly from the European Agricultural Fund for Rural Development (EAFRD) and partly from national, regional and sometimes private sources.

Pillar 2 of the CAP and regional funds also include funding for knowledge transfer and innovation, which includes support for cooperation, and can be used to implement the European Innovation Partnership for Agricultural productivity and Sustainability (EIP-AGRI) launched by the European Commission in 2012. EIP-AGRI aims to foster a competitive and sustainable agriculture and forestry sector that "achieves more from less" by bringing together innovation actors (farmers, advisors, researchers, businesses, NGOs, etc.) and connecting EIP Operational Groups and multi-actor projects, to facilitate the exchange of knowledge, expertise and good practices and to establish a dialogue between the farming and the research community.[4] Pillar 2 can be used to fund the setting up of "operational groups"; funding operational group projects (cooperation investment, knowledge transfer, advisory services); and establishing "innovation support services", e.g. to facilitate the formation of operational groups.

Horizon 2020, the EU Framework Programme for Research and Innovation, includes funding for food and agriculture research projects through the theme "Food Security, Sustainable Agriculture and Forestry, Marine, Maritime and Inland Water Research and the Bioeconomy".[5]

National implementation

The implementation of the CAP reform decisions by the Netherlands is summarised in Table 6.2, which distinguishes the financial aspects, the architecture of the new direct payments and the implementation of the second pillar (Rural development).

Table 6.2. National implementation of the CAP over 2014-20 in the Netherlands

Financial aspects	
Direct payments	As a result of the Multiannual Financial Framework (MFF)/CAP reform, direct payments for all EU27 member states were cut by 3.2% compared to the baseline (envelopes in the existing regulation prior to the reform with full phasing-in). The Netherlands contributes to the financing of the convergence of direct payments. The baseline envelope for the Netherlands for financial years 2014-20 is EUR 5 814 million. With the MFF/CAP reform (i.e. external convergence + MFF cut), Netherlands' envelope for the seven-year period amounts to EUR 5 405 million (all amounts in current prices). All abovementioned amounts are before financial discipline that will be applied to all direct payments and subject to the application of the reduction of payments as of 2016 (Article 11 of Regulation 1307/2013).
Rural development	Netherlands' envelope for 2014-20 set out in Regulation (EU) No. 1305/2013 amounts to EUR 607 million (in current prices). Compared to a situation where the 2013 envelope would have been maintained (in current prices) over the period 2014-20, this means that the decrease in the Netherlands' envelope is larger than the average decrease for EU27.
Flexibility between pillars	The Netherlands will not transfer any funds between the two pillars for calendar year 2014 (financial year 2015), but the government has announced its plan to transfer 10% of Pillar 1 direct payments to Pillar 2 as from calendar year 2015 onwards.
Architecture of the new direct payments under Pillar 1	
Active farmer	Exclusion for greenhouses. The negative list may be broadened by public authorities and semi-public organisations.
Capping and degressivity	Minimum payment is set at EUR 500 per farm. Capping by 5% for direct payments above EUR 150 000. (No further increased capping is foreseen).
Model until 2014	Historical

Table 6.2. National implementation of the CAP over 2014-20 in the Netherlands (*cont.*)

New model (since 2015)	Full convergence by 2019 (linear downwards evolution as from 2015-19) One uniform "flat rate" basic premium (EUR 270/ha) and one uniform green payment (about EUR 120/ha) For the creation of new payment entitlements in 2015 the value of 2014 entitlements will be used as a starting point. Eligible hectares: exclusion of nature grounds which are rarely (or very extensively) used for agriculture
Redistribution payment	No
Young farmers	"Maximise all foreseen possibilities" so probably up to 2%
Small farmers scheme	No
Areas under natural constraints	No
Greening	Crop diversification: no national discretion. Permanent grassland: monitoring at national level, at start with no obligations at individual level Choice of Ecological Focus Area (EFA) (landscape elements, equivalent set of agri-envi measures, sustainability certificates, possibilities for collective approach Equivalence amongst others through protein crops (as part of combinations sets).
Voluntary coupled support	"Grazing premium" for those animals (sheep and cows) grazing on the nature grounds that will be excluded as eligible hectares (see above), with a budget of EUR 3.5 million per year, i.e. less than 5% of the national envelope for direct payments..
Implementation of Pillar 2	
Programme structure	One programme with a national part and provincial parts
	Limited set of measures and operations (simplified in comparison with previous programme) + former "axis 3" measures are taken out of the new programme and will no longer be co-financed (purely state aid), new programme will focus mainly on farmers and the environment.
Focus	Innovation (precision farming) and competitiveness (insurance scheme) Young farmers (dedicated envelope of EUR 5 million per year for investments) Nature and landscape Improvement of water quality Leader

Source: Institute for Prospective Technological Studies (IPTS), https://ec.europa.eu/jrc/en/institutes/ipts.

Direct payments

As the Netherlands currently applies the historical model, payment rates per hectare vary significantly by farm specialisation (Figure 6.5).This implies that the move towards a single rate will have large redistribution effects. On average, payments in the Netherlands are EUR 430 per hectare. Farm types that receive a payment per hectare in 2014 that is greater than the average payment will receive a reduced payment, whereas farm types that receive a lower than average payment per hectare in 2014 will receive more.

For around 70% of the farms, the changes in the payments received lie between minus EUR 5 000 and plus EUR 5 000. Around 13% of the farms will be more than EUR 5 000 worse off, while approximately 17% of the farms will see their payments increase by more than EUR 5 000 (Terluin, Jager and Jongeneel, 2014).

Figure 6.5. Average farm payment per hectare, by farm type, Netherlands, 2014

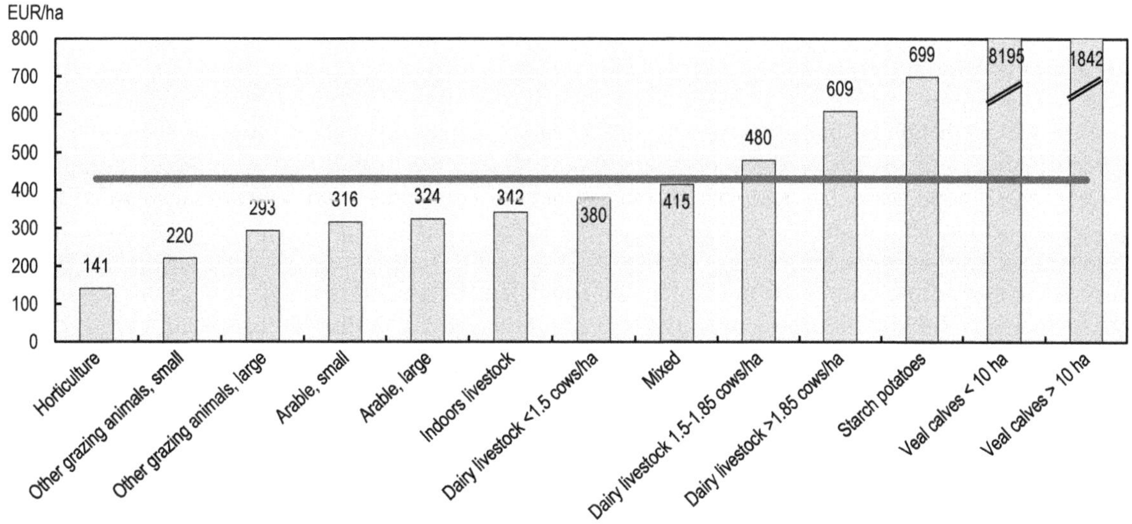

The average calculated farm payment per hectare (430 euro/ha) is indicated as a horizontal blue line in the figure.

Source: Terluin, Jager and Jongeneel (2014), *Convergentie toeslagen Eerste Pijler GLB in Nederland, 2014-2019.* http://www.wageningenur.nl/upload_mm/c/d/2/74e59394-ec27-4fdc-bac9-9bd4fb82eba4_15561_LEI_Rapport%202014-001_Jongeneel_WEB.pdf.

StatLink http://dx.doi.org/10.1787/888933251157

Rural development measures

The European Commission has approved the Dutch Rural Development Programme 2014-20 (POP3) in February 2015. In POP3, the agricultural entrepreneur has a central role to further innovation and sustainability in the agricultural sector and to improve biodiversity. The allocation of EU and national funds between objectives in 2014 is as follows:

- EUR 44 million are earmarked for strengthening innovation, sustainability and competitiveness. These include the dissemination of knowledge, demos, European Innovation Partnerships, investment, and agricultural enhancement. There is also a guarantee scheme for risky innovations and broad weather insurance managed by the Ministry of Economic Affairs.
- EUR 5 million are dedicated to supporting young farmers.
- EUR 86 million are for managing nature and landscape. This includes nature pact, agricultural nature, geese management, hydrological measures and Programmatische Aanpak Stikstof (PAS) (Nitrogen).
- EUR 25 million are to improve water quality. This includes organisation, management and innovation in this area.
- EUR 11 million are dedicated to LEADER development projects.

In terms of distribution of funds, the RDP for 2014-20 (POP3) has a stronger environmental focus: measures to promote agricultural sustainability account for 58% of the total, while investment support to improve the competitiveness of the food chain account for 21% (Table 5.1). It aims to have 6% of the agricultural land under contact to stimulate biodiversity and improve water and soil management (Agra Europe, 20 February 2015). It also supports innovative and sustainable investments, targeting 4% of all farms including those of young farmers. For the first time, some CAP funds are available for knowledge transfer and innovation (4% of the total). The focus of EU co-funded policies on innovation is stronger when ESI funds included in the Partnership Agreement for 2014-20 are considered as this thematic objective accounts for 21% of all funds (just above the mandatory minimum of 20%).[6] The programme also includes support to an all-weather insurance scheme which targets 2% of farmers.[7] In accordance with the EU regulation, other sources of funds are used for non-sectoral rural development.

General services for agriculture

In the Netherlands, most national expenditures to support agriculture[8] are for the provision of general services to the sector (88% of total support to agriculture in 2014), as support to producers comes largely from the EU budget. When accounting for EU payments to producers, but not for market price support included in the EU Producer Support Estimate (PSE), support to general services accounts for 41% of total support, which is relatively high by OECD standards. Most of support to general services to agriculture goes to green training and education (87%), and reaching 91% to innovation with the inclusion of R&D, as infrastructure investment is rarely agricultural-specific, and companies pay for most of the inspection and control costs. Government support to innovation is developed in the following Chapters.

The government also supports monitoring in the field of economics and the environment. On this basis, the parliament and the civil society are informed and the policies can be adjusted if necessary.

Policy impact

A major part of the Dutch agri-food sector has not been covered by the CAP or EU agricultural trade policy, namely horticulture and the processing of imported tropical products. The horticulture sector has been able to develop successfully into a competitive, innovative player on the world market without EU support. Strict environmental regulations may have triggered the innovative capacity of the sector (e.g. Glass Greenhouse as Source of Energy, Box 2.2).

Trade policies protecting EU markets from foreign competition and allowing free import of inputs (feedstuffs) have contributed to the development of very intensive livestock production systems, which generate high quantities of animal waste. Domestic support has concentrated on dairy and beef production systems. With higher farm receipts per unit of ouput and other investment aids, farmers have been able to invest in labour-saving and yield-increasing technologies. Elimination of coupled support and stricter pesticide or environmental regulations in some sectors has forced innovation (starch potatoes) such as, for example, the adoption of new varieties requiring fewer pesticides to reduce costs and environmental print.

Capitalisation of support into asset values has raised production costs and this has possibly slowed structural adjustment, which happened anyway. It has also imposed an extra burden on new entrants and farm transfer. In the case of milk, quotas have limited national production, and despite being freely traded within the country, they have also slowed adjustment by constraining production increases in most efficient farms. Since the phasing out of the quota started in 2007, Dutch dairy farms have expanded milk production and increased total factor productivity (Box 2.1). The Netherlands is expected to benefit further from the removal of quotas although environmental and other constraints may limit production expansion in the future.

Adaptation to policy incentives has led to high levels of total and partial productivity, and sustained productivity growth. Excess of manure, and more particularly its negative externalities on soil, water and air quality, is a major issue and requires innovative solutions to find ways to dispose of it (new usage, transport, energy).

The development of environmental regulations and targeted measures has helped reduce the environmental costs of an intensive production system. Innovation has had an important role in helping farmers pursue environmental targets, and cope with higher input costs (e.g. energy, pesticides). This is a clear case of demand-driven innovation, where co-operation with farmers on good practices is very important.

The food industry adapted to high domestic prices for raw materials by diversification in tropical products, becoming competitive in the EU market, being one step ahead in terms of sustainability. Innovation took place in the collection and processing chain to compensate for high input costs. High productivity in milk compensates partially for higher cost of raw material.

The challenge for agricultural policy is to foster the adoption of innovations that help adaptations to environmental conditions, while remaining productive and competitive, and take advantage of market opportunities. Policy should also give clear signals to the industries about future policy directions, as preparedness to policy change is crucial for the sector. Policy uncertainty and lack of credit are cited as the main obstacles to investment – and also for the innovation system as innovation requires time to develop.

Summary

- Within the framework of the EU Common agricultural policy, the Netherlands generally opts for measures that facilitate productive investment and distort the least market signals. Policy decisions are informed by the development of data bases and analytical tools to monitor progress and evaluate impact.
- A characteristic of Dutch rural development programmes under the CAP (Pillar 2) is the concentration of funds on a limited number of measures to reinforce impact. In this context, increasing emphasis is placed on innovation, sustainability and nature preservation.
- Developments in EU policies offer opportunities for Dutch agriculture. The abolition of EU dairy quotas will facilitate further investment in large-scale, innovative technologies needed to remain competitive, and improve further productivity and environmental performance, while greening is expected to increase areas covered by good environmental practices above those already covered by environmental schemes.
- It is crucial for agricultural policy to provide a long term vision for the sector, which recognizes the need to improve environmental performance while maintaining productivity growth.

Notes

1. EU Agricultural and Rural Development web site: http://ec.europa.eu/agriculture/.
2. For more information on CAP history, please see: http://ec.europa.eu/agriculture/cap-history/index_en.htm.
3. The description of the measures below is based on OECD (2011), Evaluation of Agricultural Policy Reforms in the European Union, OECD Publishing, Paris. http://dx.doi.org/10.1787/9789264112124-en.
4. For more information on the European Innovation Partnership for Agricultural productivity and Sustainability, see: http://ec.europa.eu/eip/agriculture/.
5. Horizon 2020 website: http://ec.europa.eu/programmes/horizon2020/en/h2020-section/food-security-sustainable-agriculture-and-forestry-marine-maritime-and-inland-water.
6. 2014-2020 Partnership Agreement: http://ec.europa.eu/contracts_grants/pa/partnership-agreement-nederlands-summary_en.pdf.
7. This scheme was previously funded under article 68 of Council Regulation (EC) No. 73/2009.
8. National expenditures are classified according to the OECD methodology for estimating support to agriculture.

References

OECD (2014), *Agricultural Policy Monitoring and Evaluation 2014: OECD Countries*, OECD Publishing, Paris.
DOI: http://dx.doi.org/10.1787/agr_pol-2014-en.

OECD (2011), *Evaluation of Agricultural Policy Reforms in the European Union*, OECD Publishing, Paris.
DOI: http://dx.doi.org/10.1787/9789264112124-en .

Oskam, A.J., G. Meester, and H.J. Silvis (Eds.) (2011), *EU policy for agriculture, food and rural areas*, 2nd revision, , Wageningen Academic Publishers, Wageningen.

Terluin, I.J., J.Jager and R.A. Jongeneel (2014), *Convergentie toeslagen Eerste Pijler GLB in Nederland, 2014-2019*, Wageningen, LEI Wageningen UR (University & Research centre), LEI Report 2014-001. 58 blz.; 23 fig.; 17 tab.; 3 ref.
www.wageningenur.nl/upload_mm/c/d/2/74e59394-ec27-4fdc-bac9-9bd4fb82eba4_15561_LEI_Rapport%202014-001_Jongeneel_WEB.pdf .

Chapter 7

The Dutch agricultural innovation system

This chapter describes the Agricultural Innovation System of the Netherlands and outlines recent changes. It first provides an overview of the general innovation system; describe agricultural innovation actors and their roles in the system; outlines changes in roles and changes in themes; present main policy instruments and monitoring mechanisms; and discusses views in the general public on agri-food innovation. It then describes main trends in public and private investments in R&D, mechanisms of funding and mechanisms to foster knowledge markets and networks. The next section presents an overview of policy incentives for the adoption of innovation, outlines the role of training and advisory services at farm level, and provides some information on adoption rates in primary agriculture and food processing. It concludes with reflections on recent developments. Finally, the last section outlines the participation of Dutch agricultural R&D actors in EU and international co-operation.

The statistical data for Israel are supplied by and under the responsibility of the relevant Israeli authorities. The use of such data by the OECD is without prejudice to the status of the Golan Heights, East Jerusalem and Israeli settlements in the West Bank under the terms of international law.

General innovation profile

Agriculture innovation[1] is increasingly dependent on general innovation, through developments in ICT, biotechnology and nanotechnology, but also marketing innovation and cross-sectoral innovations. A thriving innovation profile will ensure that general knowledge and specific knowledge in other fields (needed to develop and implement agriculture innovation) are available, and that economic actors and society in general share an innovation culture (OECD, 2014a).

Performance

The Netherlands has a well-performing knowledge economy relative to investments. Economic development has been based on continuous innovation and the country has achieved one of the highest GDP per capita. Gross domestic expenditures in R&D (GERD) as a percentage of GDP are, however, lower than the OECD average, with the gap further widening in the 2010s. At the same time, the numbers of triadic patent families and publications in the top journals per GDP are close to the OECD top 5 performer average (Figure 7.1).

According to OECD statistics, the Netherlands spent about 2% of GDP on R&D in 2013, which is close to the EU average. The share of GERD financed by the public sector in the Netherlands is above the OECD median (Figure 7.1). Most public R&D funding is disbursed as block funding to universities and research institutes, but the funding for projects, selected on a competitive basis, increases in importance (OECD, 2014b, Figure 9.32, Panel 4).

The contribution of Dutch business to R&D of 1.01% of GDP is comparable to the EU average (1.07%), but lower than the OECD average (1.42%), and lower than in EU countries such as Finland, Sweden, Denmark and Germany. This is explained by the composition of the economy, in which middle and high-tech sectors are relatively underrepresented. When adjusted for industry structure, the private R&D intensity in the Netherlands lies above the OECD average (Ministerie van Financiën, 2014). In fact, Dutch enterprises have strong technological capabilities and performance (OECD, 2014b).

Figure 7.1. Science and innovation in the Netherlands

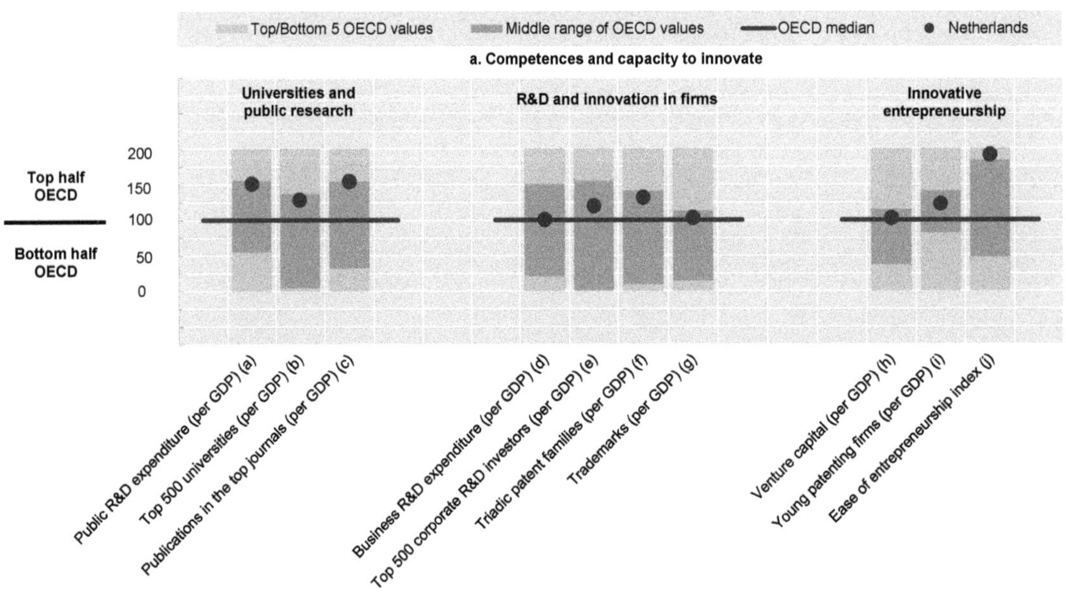

Source: OECD (2014b), *OECD Science, Technology and Industry Outlook 2014*, http://dx.doi.org/10.1787/sti_outlook-2014-en.

StatLink http://dx.doi.org/10.1787/888933251169

PCT patent applications reveal the technology advantage of the Dutch innovation system in bio- and nano-technology and ICT compared to the OECD median and the EU28, but not in environment-related technologies (OECD, 2014b, Figure 9.32).

Policy approach

Government policy towards innovation is twofold:

- It aims to provide framework conditions conducive to innovation in businesses, through streamlining regulations, improving transparency and provision of public services, providing tax incentives and improving access to finance (loans and credit guarantees).
- It focuses public investment in R&D on top sectors (Box 7.1).

The motivation for the top sector policy introduced in 2011 was to concentrate scarce public funds in export-oriented sectors facing increasing competition from emerging economies, where innovation would help maintain international competitiveness (OECD, 2014c). One original objective was to leverage business-sector R&D and increase the applicability of public research. Another one was to increase policy co-ordination and co-operation between innovation actors. In this area, the top sectors build on the work of the Technological Top Institutes, which were established to bridge the gap between research being done at the universities not reaching the private sector, and on pre-existing networks.

Box 7.1. Top sector policy

In 2011, the Dutch government introduced a new Research and Development strategy across the whole economy: the top sector policy (*topsectorenbeleid*).

As earlier R&D strategies, this new policy maintains the same demand-driven approach. It starts with the observation that although public financing of R&D has a high priority, private investments by companies and firms lag behind. It concentrates public funds on key sectoral areas (*sleutelgebieden*). The top sector approach involves public-private partnerships (PPPs) by sector, to facilitate co-ordination and increase the impact of public funds on economic performance. Public funds have to be matched with an equivalent contribution from the private sector (50-50), which can be in kind (access to facilities) or financial, in which case it can benefit from public support (investment or tax rebates).

Nine key sectors have been identified with strong market positions. They are capital intensive and particularly knowledge intensive sectors, and accounted for over 80% of business R&D in 2011 (96% in 2010), 55% of exports but under 30% of value-added and employment (OECD, 2014c). They form a significant part of the Dutch economy, with high productivity rates (35% above average), a strong position in international markets, and thereby contribute significantly to economic growth.

Investment in R&D is one of the key elements to achieve higher productivity and economic prosperity. In the top sector policy the business world sets the agenda for R&D investments in its field. The government does not make its own proposals for the sectors, but invites businesses and scientists to draw up action plans, which serve as a base to develop concrete lines of actions. More and better public-private partnerships are expected to increase the innovative power of businesses, thereby generate new products and services to face the grand challenges of the future.

The top sector policy aims to promote closer co-operation between knowledge institutes, public authorities and business. All top sectors have a human capital agenda meant to strengthen the linkages between education institutes (from vocational training to university) in order to meet the needs of the sector itself.

Each top sector has created one or more top consortia (TKI) for knowledge and innovation where entrepreneurs and researchers work together in innovative products and concepts. Three strategic goals concerning the knowledge infrastructure were defined:

- More co-operation between knowledge institutes, business and government to strengthen the international strength of the sector concerned.
- Facilitating public-private partnership for R&D.
- Giving a structural financial base to public private partnerships in the knowledge system.

Every TKI has a board with members from all three parties; government, business and knowledge institutes. Programming is done by calls to tender leading to a number of identifiable projects since 2012. The TKI allowance provides public co-funding, amounting to 25% of the private funding (*TKI-toeslag*). In order to stimulate SMEs to participate the public co-funding is 40% for the first EUR 20 000 per project. Arrangements between government and TKIs are written in the so-called innovation contracts (*innovatiecontracten*).

Source: Investing in top sectors | Entrepreneurship and innovation | Government.nl.

After three years of implementation, it is still early to evaluate the full potential of the top sector policy. However, experts have identified opportunities, trade-offs and risks of the top sectors. The discussion has been summarised in previous OECD reports (OECD, 2012, 2014c). Box 7.2 contains the considerations reported in OECD (2014c, pp. 187-188)

Box 7.2. Opportunities, trade-offs and risks of the top sectors

In the discussion of the opportunities, trade-offs and risks of the top sectors contained in OECD (2014c), experts noted that the "emphasis on sectors of strength (with implications for the representation of incumbents versus challengers and the absence of a search for new niches, makes it somewhat less dynamic than other forms of modern industrial policy." In other words, the approach supports existing firms and innovation activities, but there is a risk that it does not facilitate the emergence of new innovative firms and innovation areas.

As an approach to innovation policy, "it increases the likelihood that bottlenecks outside the traditionally narrow remit of innovation policy will be identified, and that sufficient attention and resources will be diverted to tackle them. It also introduces novel forms of governance, through the involvement of stakeholders not only in policy formulation but also in policy delivery and implementation."

"An OECD Economic Survey (OECD, 2012) expressed concerns regarding its possible impacts on the wider business sector. Many of the concerns raised are familiar to almost any debate about selective industrial policy: the government has incomplete information to pick future winners, the process runs the risk of capture by well-organised interests, and co-ordination processes can be bureaucratic and inefficient. Other arguments take issue with specific aspects of the top sectors approach. Among them are claims that a sectoral approach does not take account of global value chains, that it risks diverting resources from horizontal policies related to education, fundamental research and the provision of public goods more generally, and that the gains from co-operation between business and government can be overrated.

Moreover, it has been argued that some aspects of the approach may undermine its own objectives. There are concerns with the selection of sectors, the alleged tendency to favour incumbents at the expense of challengers, the alleged lack of sufficient "critical" mass, and the balance between small and large firms. A common objection to the current choice of sectors is that it is backward- rather than forward-looking, especially in terms of emerging social challenges. Another objection is that it is predominantly technology-oriented, with insufficient attention to non-technological innovation and the role of social sciences.

It would also be important to take account of other possible risks. A key issue is the complication of principal-agent dynamics when the government is part of the top teams and also has a mission to regulate markets. The design of an appropriate governance framework must be based on the understanding that the interests of government and business do not always overlap (even when the economy is concerned) and particularly if those involved are mostly large firms in certain sectors. The diversion of government policy attention and regulatory interventions in some sectors but not in others can compromise the coherence of the policy mix and of policy delivery. Of course there is always potential to improve co-ordination between government ministries and therefore lessen the regulatory burden faced by the sectors concerned. However, the Netherlands' good position in indicators of the business climate suggests that the magnitude of the likely improvements is small, and perhaps smaller than the risk of regulatory capture and of increased heterogeneity of regulation across sectors. Provided such pitfalls are avoided, sector-specific regulation can make sense in the interest of improving responsiveness to emerging technologies and to social challenges."

Source: OECD (2014c), *OECD Reviews of Innovation Policy: Netherlands 2014*, DOI: http://dx.doi.org/10.1787/9789264213159-enhttp://www.oecd-ilibrary.org/science-and-technology/oecd-reviews-of-innovation-policy-netherlands-2014_9789264213159-en?citeformat=ris.

In a recent report on the status of the top sectors in 2014, the Advisory Council for Science, Technology and Innovation (*Adviesraad voor wetenschap, technologie en innovatie,* AWTI) suggests the following improvements to ensure a shared vision; improve governance and forge links between all innovation actors:

- The vision of top sectors, which is to strengthen the Dutch economy and to solve societal challenges, which are often related to economic performance by bolstering public-private partnerships in a number of key areas, should continue to be communicated clearly and unambiguously along with the secondary objectives, especially among mainstream, innovative SMEs (including start-ups and SME growth-drivers), universities of applied sciences, regional authorities and line ministries; and engage these actors in a dialogue to create a shared and well-articulated vision.
- The government, in this case the Ministry of Economic Affairs, must take on a proactive and facilitating role with the aim of reducing the administrative burden experienced in the field.

- The range of innovation instruments should be re-examined. Should additional funding become available from the national government, the budget for the MIT scheme and/or the TKI scheme can be significantly increased. If no additional funding becomes available, then the TKI scheme should be simplified by interpreting the equity principle behind the scheme more flexibly.
- Substantive visions and strategic agendas of the top sectors are needed with regard to the societal challenges as these challenges can only be effectively addressed by taking a cross-sectoral approach. This will facilitate the inclusion of all relevant players in new cross-sectoral potential areas for collaboration, and participation in international networks working on the Horizon 2020 societal challenges.
- The addition of new players in top sector governance should be considered, for example by including a member to the top team from the universities of applied sciences, regional authorities and/or start-ups and SME growth-drivers; and professors from universities of applied sciences in the TKI programme boards.
- Additional instruments, other than fiscal or revolving instruments, will be needed to facilitate collaboration between different types of actors, such as large companies and SMEs, joint projects between universities of applied sciences, senior secondary vocational education, research universities and companies. This will be difficult to achieve using the existing range of instruments, and it is also at odds with the wish of both the national and regional governments to implement a range of instruments with a revolving character wherever possible. (AWTI, 2014).

General governance of innovation

The main actors in the design and definition of Dutch innovation policies are the Ministry of Economic Affairs (EZ), which is responsible for promoting competitiveness, entrepreneurship and innovation; and the Ministry of Education, Culture and Science, which is responsible for defining strategies and policies for public-sector education and research (OECD, 2014c). Both ministries co-ordinate the science policy agenda of the national government and contribute to the definition of international science policy at the EU level and beyond. As the ministry in charge of agriculture, the Ministry of Economic Affairs also funds agri-food research and green education.

A number of councils and advisory bodies to government and policy bodies represent multiple actors of the innovation system, including, in some cases, the business sector. They include for example the Advisory Council for Science and Technology Policy (AWT), the Royal academy of Arts and Sciences (KNAW) and the Netherlands Scientific Council for Government Policy (WRR). For a complete list and description of research and advisory councils to the Dutch Government, see Section 5.2 of OECD (2014c, pp. 176-77).

The government plays a central role in R&D funding. The Netherlands Organisation for Scientific Research (NWO) is responsible for distributing funds from the Ministry of Education, which are for fundamental research. The Netherlands Enterprise Agency (RVO), which is part of the Ministry of Economic Affairs, is responsible for paying investment support and tax rebates for R&D. (OECD, 2014c).

Between 2004 and 2010 agenda setting was the role of the Innovation Platform, which had high-level representation from government (the prime minister and ministers of Economic Affairs and of Education and Science), business, knowledge institutes and independent experts (Boekholt and den Hertog, 2005). At present, a considerable amount of research system co-ordination takes place in the context of the top sectors, which rely on co-ordination of the different communities (notably academia, business organisations and government), represented in the top teams.

The Netherlands has a strong evaluation culture at all levels. Evaluation is widespread and taking place at various levels: research groups, projects, programmes, policy and organisations. Evaluation results are effectively used to improve policy instruments. So far, however, this experience in evaluation has not been applied to system-level evaluation and social cost-benefit analysis (OECD, 2014c).

The agri-food sector is fully part of the top sector policy and follows the same rules for priority setting and evaluation.

Actors, institutions and their roles in an agricultural innovation system in transition

Agricultural innovation systems involve a wide range of actors who enable, guide, fund, perform, implement, inform and facilitate innovation. The key players include policy-makers, researchers, teachers, advisors, farmers, private companies and consumers. They are commonly categorised as government, research, industry, academia, other organisations, such as non-profit organisations, and markets (OECD, 2014a).

The term "golden triangle" (in Dutch; *gouden driehoek*) is a concept used in the Netherlands to describe the traditional productive relationships between business (firms and farmers), government and knowledge institutes to stimulate innovation. The Dutch agricultural innovation system is characterised by its complexity and a large number of institutions interrelating the organisations of the triangle. As shown below, the triangle is still visible but the number of interrelated and cross sectional connections is numerous.

Knowledge institutes

There are two basic structures of agricultural knowledge institutes: the educational system and the research institutes (applied and more fundamental). A central actor in both is the Wageningen University and Research Centre (Wageningen UR). The university is for bachelor and master's courses and for PhDs. At Utrecht University there is a faculty of veterinary medicine.

Besides Wageningen University education there are 10 000 students enrolled in agricultural education (primarily bachelor level) at universities of applied sciences across the country. At a lower level there are specialised schools for vocational training at MBO-level (over 16 years) and VMBO-level (12-16 years). See Chapter 5 for more information on the green education system.

Wageningen UR consists of a research university with chair groups for academic education and basic research and in total nine specialised research institutes for applied research (DLOs). Institutes like "Plant Research International", "Livestock research", "Imares" (marine ecology and fisheries research), "Rikilt" (food safety) and "LEI" (agricultural economic research) all contribute to the body of knowledge of the agricultural innovation system.

Wageningen University is one of the 14 publicly financed universities in the Netherlands organised in the association of universities: VSNU. Wageningen UR was formed in 1997 to integrate the (formerly governmental) DLO institutes and the Wageningen University because of the considerable overlap and unnecessary competition between the two (Peper, 1996). The process for this integration lasted almost a decade, and only involved the institutes that were under governance of the former Ministry of Agriculture. The other institutes which were under the governance of other ministries (Netherlands Organisation for Applied Scientific Research, TNO, under the Ministry of Education and Science;[2] the National Institute for Public Health and Environment, RIVM, under the Ministry of Health, Welfare and Sport; and the Faculty of Veterinary Medicine under the Ministry of Education and Science) were not included in this reorganisation.

Apart from Wageningen UR there are a number of research institutes and departments; and government based, sector based and company based research facilities.

- The Institute of Veterinary Research (IVR) of Utrecht University.

- The Netherlands Organisation for Applied Scientific Research (TNO), a non-profit company (3 900 employees). A division of TNO is specialised on food safety, healthy food and innovative food concepts. It is under the supervision of the Minister of Economic Affairs.
- The National institute for public health and environment (RIVM) a non-profit institute with 1 500 employees with a Centre for Nutrition, Prevention and Health Services. This institute is part of the knowledge infrastructure of the Minister of Health, Welfare and Sport.
- The Technological Top Institutes (TTI) "Green Genetics" and, "Food and Nutrition"; and the Centre for Biosystems Genomics.[3] Government institutional support for these institutes is terminated. The Top Consortia for Knowledge and Innovation (TKIs) (Section 7.6) have integrated the knowledge and network function of the TTIs.
- A number of private companies with research and consultancy functions for the agri-food sector:
- Nizo (200 employees), specialised in applied research for food companies (originated from the dairy industry).
- The Institute of Sugar Beet Research (IRS), specialised in research and knowledge dissemination for the sugar production sector.
- The Louis Bolk Institute, specialised in sustainable agriculture (50 employees).
- DLVs (*Dienst Landbouw Voorlichting*), the former public extension services, are organised in four consultancy firms: DLV Plant-GreenQ on crop and horticulture (175 consultants) and DLV Dier on animal husbandry (120 consultants), Aequator Groen & Ruimte BV, on environmental and rural development, and DLV Agriconsult BV.[4]
- A number of other private firms for consultancy and research such as Cropeye, Zatedec, Soilandmore, Keygene, CLM, HLBBV, CGD Deventer, Schothorst, and Top BV.
- Company and co-operative based research plants like Unilever, FrieslandCampina, Danone (Utrecht science park). There are 12 out of the global top 40 food and beverage companies that have R&D facilities in the Netherlands.

Agricultural innovation institutions have been stable between the formation of Wageningen UR in 1997 and the introduction of a new R&D strategy, the top sector policy, in 2011 (Box 7.1). Within Wageningen UR there have been a number of developments in terms of the research institutes and the organisation of co-operation between institutes and the university.

The Food valley and the Seed valley have already been mentioned in the context of infrastructure and rural development (Box 5.2). Food Valley NL is a public-private cluster network organisation of food companies and research companies and institutes launched in 2004. Overall, it includes 15 000 food scientists and engineers in a radius of 50 km from Wageningen. Seed Valley Enkhuizen as a concept was introduced shortly thereafter in 2007. This valley is world leader in seed products and propagation materials.

Government

The Government plays a role in the governance of the agricultural innovation system, by setting the policy, monitoring the implementation of programmes and evaluating policies and institutions. Monitoring and evaluation mechanisms are the same as for all policies and institutions in the Netherlands. There is no overall mechanism, however, to evaluate the performance of the whole agricultural innovation system. The government provides funding for R&D and education, in public and private institutions, and innovation support to businesses.

The rationale for government investment in R&D is market failures: the private sector tends to underinvest because agricultural research is often long term, large scale and risky. Moreover, the organisation and support of the knowledge and innovation system is one of the fundamental instruments of the government to support the transition towards a more sustainable agriculture, which is not necessarily a priority for the private sector as the price of environmental outcomes is often undervalued by markets. The relatively small-scale and diversity of primary agriculture also means that the sector does not have the capacity to invest in R&D unless it gets organised and pools its capacity with that of agri-food companies.

In 2015 the government organisation responsible for the agriculture and agro-food sector is housed at the Ministry of Economic Affairs. One of the four directorates within the ministry is the DG Agro and Nature consisting of five departments; Agro and Nature Knowledge being one of them. The Ministry of Economic Affairs sets out the specific and general policy towards the sector on the knowledge infrastructure. Until 2010 with the formation of the Rutte 1 cabinet, agricultural policy was governed by a distinct Ministry of Agriculture, Nature and Food quality (LNV).

The agriculture education system although institutionally distinct from the other education institutions follows the integrated education laws set out on the responsibility of both ministers (Education and Economic Affairs). The Minister of Economic Affairs has the direct budgetary responsibility for Wageningen University. There is another important governmental link with the Ministry of Health, Welfare and Sport. One of the knowledge institutes relevant for the agricultural innovation system is situated within the RIVM, which is an institute of this ministry.

The public inspectorate (The Netherlands Food and Consumer Product Safety Authority, NVWA) is in charge of controlling the sector for food safety and animal welfare. Its activities are performed under the supervision of the Ministry of Economic Affairs with close links to the Ministry of Health, Welfare and Sport.

Apart from national policy, regional and local policy (like the municipalities of Wageningen and Enkhuizen) has a policy influence on the agricultural innovation system, among others through the Food Valley initiative.

The European Union plays a growing role in the orientation and financing of research collaboration between EU member states, including in the agri-food area. But the EU and national or regional agendas are not necessarily aligned.[5] In the EU Horizon 2020 budget EUR 3.9 billion are earmarked for research in "food security, sustainable agriculture, marine and maritime research and the bio economy". Developments in EU research priorities for the sector, from production oriented to green economy, reflect changing priorities for agricultural policy.

The food and agriculture sector

The agri-food sector is prominent in the Netherlands' economy, having a 10% share in the GDP and almost a 20% share in the export capacity. Farmers, the food industry and the supplying industries (like greenhouse construction and energy management) are the primary actors in innovation. They form a diverse landscape of organisation described in Chapter 2. As mentioned above, a number of leading (inter)national food companies have research facilities in the country.

The agricultural sector itself is organised through LTO Nederland (Dutch federation of agriculture and horticulture) with 15 sub-sectoral divisions. LTO has its own consultants advising entrepreneurs in the sector on a number of topics concerning production, specialisation, and business opportunities.

The Netherlands used to have a number of commodity boards (*productschappen*) for the agricultural sector and specific branches within the sector, eleven in total. Those were public institutes financed by the firms in the sector. They used to have an advisory position with the government, and financed and coordinated the applied research projects in the sector. These commodity boards were abolished in 2014 and their role in the organisation and funding of innovation has not been replaced (Section 3.2).

Intermediates

It is practically impossible to give a full and detailed description of all the intermediate organisations in the system. In the literature on knowledge policy for the agro-food sector, the system has been described as the OVO system, consisting of the linear sequence of knowledge production (*onderzoek*) through, communication and extension (*voorlichting*) to education (*onderwijs*). All government initiatives in the sequence were considered to be key to the success of the sector.

Today there the government provides a majority of public funding for agricultural research and education but extension services have been privatised and the former *Dienst Landbouw Voorlichting* (DLV) has been transformed into four independent private consultancy firms providing tailor-made advice to clients in the sector on a broad field of subjects (technical, economic, managerial, construction, and environmental).

The EU policy on agricultural and rural development has asked member states to develop a Farm Advisory System (FAS), for which the Ministry of Economic Affairs is the responsible authority. The FAS in the Netherlands consists (after an accreditation procedure) of 41 private advisory firms in the Netherlands.

There is an elaborate network of (informal) farmer study groups based on the principle of farmers' solidarity and co-operation, and mutual interest in sectorial or geographical proximity. They are recognised and valued by both the policy and research domains of the system and are of vital importance to improve knowledge and innovation.

Sector council and innovation network

An agro-food Innovation Network (*Innovatienetwerk: Groene Ruimte en Agrocluster*) has been in place since 2001. Until 2014 the Innovation Network played the role of an intermediary between the research activities and institutes on the one hand, and the ministry on the other hand. Since 2015 the Network has become part of the Ministry of Economic Affairs and the department of Agro and Nature Knowledge. The primary focus of the network is to transform new concepts of production in the sector towards the first stage of the development of crossing border innovations. The very essence of these innovations is their character of game changers based on a diversity of network agencies. So the network is active in the very early stages of development of new ideas, usually long before the actual application in business.

The Innovation Network is a project-based organisation: in 2013 it carried out 70 individual projects covering a specified number of topics (Innovatienetwerk, 2013). The actual three topics are: 1) Agriculture horticulture and agro business; 2) Food; and 3) Nature landscape and space. The network has a total annual budget of EUR 2.5 million.

For the sea fishing sector there was an innovation network (*Visserij Innovatie Platform*) in place from 2006 to 2010, which ended with the three-year programme supporting its activities, as planned. Its activities are continued by the so-called National Platform Blueports and the regional Blueports (www.blueportal.nl).

Changes in roles

The Dutch agricultural sector has become more and more knowledge-intensive and has achieved a high level of productivity, maintaining thus its position in the world market for agricultural products in the last decades. Much is due to a continuous investment in innovation strategies and knowledge production and dissemination. The sector has benefitted from intensive co-operation between public research and education institutes, farmers, and agro-food firms and co-operatives. Entrepreneurs in the sector have become more and more specialised professionals, who have had higher education and training themselves. As a result, the whole agricultural innovation system has become more complex and the roles of actors have changed.

The government has a long history in a pro-active role of promoting both the creation and diffusion of knowledge and innovation in the sector. State involvement in the agricultural innovation system should be understood as the correction of market failures, which result in the relatively modest investments of private parties in the knowledge system observed in the past, and which made government involvement of vital importance to bring to the sector the innovation needed for its development (De Haas, 2013, pp. 75). According to a recent report on the history of state involvement in the Dutch agro-food sector (De Haas, 2013), the government has been quite successful in strengthening innovation and competitiveness in the sector.

Until the 1990s, the government supported the OVO-system (research, extension, education), which had been successful in the post-war decades. This linear system has been transformed significantly and evolves continuously. From a stable and very successful system with well-defined actors and stable relations, it has become a more dynamic and diversified system with actors having a complex cluster of tasks and functions. The decentralisation and privatisation of significant parts of the knowledge infrastructure in the 1990s has resulted in the participation of more diverse actors, and the creation of new relationships. But promoting innovations in the firms of the sector has always been the objective of government involvement in this area.

Schools (from secondary education to Wageningen University) participate increasingly in networks with research and entrepreneurs and some of them become entrepreneurs themselves (as the spin-offs of Wageningen University). Institutes known for their knowledge transfer join the knowledge production arena as agri-food firms do (see for example the case of FrieslandCampina and Danone).

The role of the government has changed from being a pro-active actor in the knowledge production and dissemination in the 1980s and the early 1990s to being an actor in the "golden Triangle" of government, private firms/NGOs, and public research and education. Collaboration between public andprivate actors is the main principle for the functioning of the Dutch agricultural innovation system with a large and diverse landscape of intermediates to keep the parts connected. In the 1990s changes in the system have occurred in the light of new public management (client orientation, separation of policy and operation functions, competition) which occurred in other countries at the same time, as Roseboom and Rutten (1998) have shown. Another reason for a changing role of government lies in the sector itself. The justification for state involvement in the agricultural innovation system has been to address the system failure of insufficient private investments in applied knowledge. The main cause of this market failure was the composition of the sector with a large number of relatively small firms. This has changed over the last decades, which makes the reconsideration of the role of government logical. The government role remains crucial, however, in areas prone to market failures because of high uncertainties, such as pre-competitive research, or because of clear public good characteristics, such as environmental protection.

Extension services probably underwent the most dramatic changes. It transformed from a public institution with a supply driven philosophy and institutional public financing to an entirely private organisation with demand-activities and related funding. Privatisation took place in 1998. The public funding disappeared and the organisation had to face competent competitors for their services throughout the entire spectrum of the agricultural innovation system. Smaller firms with limited innovative capacity of their own and a strong dependence on knowledge on production and marketing developed for the whole sector have transformed into firms larger in scale, more knowledge-intensive and with stronger innovative power that have become better at articulating their demand for knowledge.

Privatisation resulted in the transition of knowledge as a public good to knowledge as a marketable product on a worldwide market. Knowledge itself has become a product in a market of supply and demand instead of a widely available source for all entrepreneurs (large and small, starters and established, wealthy and impecunious) (Caggiano, 2014).

This states the question of free and open access to knowledge and innovative concepts in the system. Knowledge being a marketable product which is owned and sold has changed the roles of all participants in the system; and may very well produce new winners and losers. Firms and larger agricultural producers may very well be in the position to pay for advice and obtain the insights and ideas they need to maintain or expand their position. Smaller firms, however, may not have enough money to afford that kind of advice and therefore will be deprived of the access to innovative knowledge (Caggiano, 2014). This prominent role of knowledge in the system which is the cause of success in recent decades in turn may be a threat for the innovation capacity of the system knowing that many innovations come from smaller firms.

The agricultural innovation system is under pressure, among others because of cuts in public funding due to austerity measures of the central government and the abolishment of statutory product boards in the agricultural sector and related levies, which used to fund R&D and innovation and have not been replaced. Consolidation in primary agriculture and agri-food industry, as well as the changing power relations in the supply chains are another element that should be taken into account, when considering changes in roles. Budget cuts and changing roles take place in a context where the scope of food, agriculture and horticulture broadens to new areas and challenges with strong public good aspects and requiring multidisciplinary approaches and cross-sector co-operation, as explained below.

Changes in themes

With the changing role of government and more demand-driven approach to research priorities many changes came in the research itself. The linear concept of knowledge flow (from university through extension to farmer and industry) was replaced by a governance model where demand and supply of research was more interwoven and dynamic (knowledge circulation). Alongside governance changes, other changes have become manifest:

- Dynamics of production and food security and the care for environment, sustainability, animal welfare and climate change. In the post war decades the main goal of agricultural research was towards volume and quality of production. The themes that were central in the agricultural innovation system were according to this goal. With changing policy in the 1980s, the focus was more on the sustainability, environmental care and animal welfare. So, other topics came into the research agendas. The main concern was how to maintain (and enlarge) the production capacity with respect to and in accordance to the environment. Recent incidents show that food safety is a constant concern for the sector and the governance, which influences the research agendas today. Sustainability as well has become a driving force for the development of new knowledge and production systems like greenhouse producing energy (instead of consuming it) and water efficient systems
- Internationalisation of both production and knowledge became more apparent. The growing internationalisation of the agricultural innovation system is particularly apparent at Wageningen University where students, who come from 100 countries all over the globe (Spiertz and Kropff, 2011), and academics are of an international character. But internationalisation implies more than increased mobility of innovation system actors. Research is not only directed towards production in the Netherlands but also includes knowledge for production all over the world. Knowledge in the Netherlands has become an export product for governments and firms globally.
- The complexity of agricultural and food systems calls for interdisciplinary analyses based on a wide variety of scientific and technological disciplines. Both the knowledge of large ecosystems and the specific knowledge of horticulture systems are required to maintain the prominent position of the sector. As a result, the knowledge systems are not only about plants and animals but also about construction and maintaining energy and water efficient greenhouses, microbiology, informatics and business economics (to name a few). All these disciplines play a role in the system and they are more and more integrated in object oriented research and innovation. This linkage of disciplines as ICT, construction, energy and health are essential for innovation because in this integration lies the real challenge of innovation itself (Roosenboom and Rutten, 1998, page 1122).

R&D policy: Strategic framework, instruments and monitoring

The Dutch government supports the agricultural innovation system by promoting public-private partnership in knowledge exchange in top sectors dedicated to agri-food and horticulture, offering tax facilities for entrepreneurs, financing R&D investments in publicly-funded education at all levels, and by funding research institutes and initiatives. A central actor is the Ministry of Economic Affairs.

Top sectors for food, agriculture, and horticulture

Implementation

Agriculture-related R&D policy is fully integrated into the broad top sector strategy (Box 7.1). Two top sectors "Agri-Food" and "Horticulture and Propagation Materials" are dedicated specifically to food and agriculture, and to horticulture respectively. They are integrated through the participation of common agricultural knowledge institutions. Other top sectors "life sciences", "energy", "water" and "high tech systems and materials" are relevant for cross-over boundaries to the agricultural innovation system.

The agriculture and food sector is, with an annual EUR 73 billion turnover, one of the most important sectors of the Dutch economy. Between 2004 and 2011 its exports doubled. There is one top consortium with over 100 businesses and knowledge institutes on a variety of subjects. The **Topinstitute for Food and Nutrition (TIFN)** is a public-private partnerships central to this top sector, with the participation of all the major players in the field and a common agenda with selected themes like nutrition and health, bio-ingredients and functionality, food chain sustainability and dynamics (See the TIFN website for more information).

In line with government policy concerning the Technological Top Institutes (TTIs), the top sector "Agri-Food" is working on a grand design to integrate the co-operation between the TNO, DLO and TIFN. There are cross-theme platforms for bacterial genomics, nutrigenomics and knowledge management.

The top sector "Horticulture and Propagation Materials" is a sector the Netherlands are famous for all over the world. It is a knowledge intensive sector focusing on solutions for food security and food safety with a worldwide exposure. There is a strong focus in finding means for producing more with less energy, water and space. There are two topconsortia for two distinct subsectors: 1) horticulture and 2) propagation materials (seeds, cuttings, bulbs).

As for all top sectors, Top Consortia for Knowledge and Innovation (TKIs) (three) have been set-up for the agri-food, and the horticulture and propagation materials sectors. Within top sectors, the relationship between agri and food (processing) has become closer, addressing various aspects including societal challenges.

With the implementation of the top sector policy private companies, knowledge institutes and the government together are responsible for the agenda setting for innovation. There are a lot of cross-sectoral connections between firms in agriculture/horticulture and other (top) sectors, as Roosenboom and Rutten (1998) predicted back in 1998.

Monitoring

The basis for the implementation of innovation activities is the Innovation Contract. Each Top Sector draws up an Innovation Contract, in which researchers, entrepreneurs and the governments (represented in the so-called Top Team) agree on measures (mix of fundamental research, applied research, valorisation), plans to develop innovative products and services, and financial contributions.

The TKIs have laid a base for monitoring the results through the innovation contracts with central government on one side and with the projects on the other side; projects within the top sectors are obliged to report on their activities and results on a yearly basis. The individual TKI's report on their websites about their activities as was done in the report by the Ministry of Economic Affairs on their enterprise policy – but it is too early to tell what the results are in terms of the goals set out earlier.

There are two general monitoring activities; both initiated by the ministry and its service units (CBS). They are about the top sectors themselves but not necessarily about the projects under the innovation contracts.

- CBS has a monitor of top sectors. The first edition was reported in 2012 (CBS Monitor, 2012). It presents the characteristics of the sector itself and reports on the number and size of companies, R&D investments, and number of personnel for example. One of the findings is that the private sector spends a total amount of EUR 400 million on R&D for food and agriculture. More than 80% comes from the industrial food companies, which do most of their research in their own facilities.
- Panteia follows a panel of firms in the top sectors and reports results on the basis of telephonic surveys. The results of the first two are interesting, also for the agricultural innovation system. One of the outcomes is that in order to develop new products and services, a significant number of companies collaborated with knowledge institutes, especially with universities (probably with Wageningen UR): 78% of the companies seek co-operation and knowledge institutes are the most wanted partners. The number of companies that sought co-operation with the TKIs was rather low (until summer 2013). Only 1% of the companies in agro-food and 13% in horticulture did so. Although the distance between both surveys was rather short (autumn 2012 to spring 2013) a significant growth in the number of companies that do co-operate was found. Those co-operating with knowledge institutes more than doubled. In terms of output parameters (new products, new processes) there were hardly any changes. The only exception was a slightly larger number of companies within horticulture that introduced new products in the last year.

Both monitors give information about the top sectors and the developments of companies. But since a very small proportion of the companies is involved in TKI projects their significance for assessing the results of this policy may be limited.

Other direct support to private R&D investment

The SME Innovation Stimulation Top sectors (MIT) scheme was introduced in 2013 to promote the participation of SMEs in top sector activities. It funds participation in R&D collaborative projects, feasibility studies, innovation vouchers, hiring of experts, networking and coaching (OECD, 2014b). Table 7.4 provides an overview of the relative weight of different sources of public funding for private investment in R&D and innovation. The extent to which they are used by agri-food and horticulture firms is discussed in Chapter 9.

Fiscal incentives to R&D

Instruments

The most central instruments in promoting R&D activities in private firms are fiscal arrangements for entrepreneurs. These are general arrangements which come under the Ministry of Economic Affairs and also apply to the food, agricultural and horticulture sector.

Fiscal incentives to private R&D include three main instruments, which are general in scope to SMEs, self-employed and multinationals in all sectors of the economy:

- The R&D payroll tax allowance (**WBSO**) offers since 1994 a contribution towards the wage costs of employees directly involved in R&D.[6] This provision is not relevant for the primary agricultural sector, which does not have such employees.
- The R&D Allowance (**RDA**), introduced in 2012, is directed towards deducting R&D investments in equipment and exploitation costs.

- The **Innovation Box** (*Innovatiebox*), introduced in 2010 as a successor of the previous "*Octrooibox*" operating since 2007, is directed towards lower taxes for benefits from WBSO projects and patents.

Monitoring

The activities and results are monitored by the Netherlands Enterprise Agency who administers the acts. The number of firms applying for these instruments and the money involved is rising over time as discussed in Chapter 8.

A Panteia study covering the period 2006-10 provides information on developments in tax benefits and number of firms benefiting from WSO and RDA arrangements and evaluates their impact on innovation and productivity (EIM, 2012). Main results show a significant positive effect of WBSO incentives on large firms in terms of the increase of R&D expenditures. For smaller firms WBSO provided an incentive to start R&D activities in order to be able to innovate. Econometric estimates find that on average, each euro of WBSO tax reduction was accompanied by 1.77 euros of private R&D. However, 55% of private R&D would likely have taken place regardless of the WBSO. In addition, decreasing returns were found as the average share of tax reduction increases, leading the government to reduce the rate of the tax benefit (OECD, 2014c). There is also a positive effect of WBSO on innovation (new products) and labour productivity. The administrative burden of this policy for both companies and the government agencies is relatively modest. Empirical evidence for the effect of the WBSO on innovation since 2010 has yet to be examined. The extent to which these tax benefits affect agri-food industries is mentioned in Chapter 8. An evaluation of the Innovation Box is due in 2015, but it is difficult to say at this stage whether it will contain information at the sector level.

Policy and monitoring of public education and research at Wageningen University and the agricultural universities for applied sciences

The policy towards the green education system follows the laws of the educational subsectors (Section 5.3). All laws are signed by both ministers (Education, Culture and Science, and Economic Affairs). Wageningen UR and other educational institutes have a governance code of their own, following the governance codes of the federation of universities and the universities of applied science, respectively.

Concerning the higher educational system the two umbrella ministries set up a strategic agenda for the future (2012-16) followed by performance agreements per institute. The one for Wageningen UR was signed in late 2012. The federation of universities (VSNU) and the union of universities for applied sciences (HBO) presented an annual report for the performance agreements concerning the year 2013. A special review committee has been set up to monitor the results of research and education over the period 2013-16.

Box 7.3 summarises the findings of various sources of information on the research and education performance of green education institutions. Some are also mentioned in Section 5.3.

Policy and monitoring the Wageningen UR-DLO institutes and TNO-food and nutrition

The DLO institutes and TNO food and nutrition, which are vital parts of the agricultural innovation system, are independent organisations, funded for a large part by government funds from the Ministry of Economic Affairs. As explained earlier the DLO institutes form part of Wageningen UR and the TNO food and nutrition is part of the larger TNO organisation.

Governance of the DLO-institutes originally was with the Ministry of Economic Affairs, as these institutes were governmental organisations until the formation of Wageningen UR in 1997. Since then there have been major changes both in terms of governance and funding. The Ministry currently finances 43% of the research budget of DLO, but there are great differences between the institutes (Wageningen UR, 2014). Some of them have been successful in obtaining private funding, others because of their orientation have been able to do this to a lesser extent. Their role in the top sectors is prominent. In the above mentioned top sectors of "Agri-Food" and "Horticulture and Propagation Materials" they have a prominent role with a budget of EUR 51 million in 2014 (*Concept onderzoeksprogramma Topsectoren* WUR DLO 2014). The DLO institutes also look for opportunities in other top sectors (Imares in the top sector "water and energy").

There is no general report on the governance of the DLO institutes. As part of Wageningen UR they are subject to the governance code of Wageningen UR. Every institute has its own management, set of clients and contract partners. Some DLO-institutes like Rikilt work with a management contract with the ministry including the statutory duties these institutes have. The Wageningen UR makes a strategic plan for a four-year period, which is submitted to the minister. There is a review of each institute by an independent review committee on a five-year sequence. The quality of the research activities is evaluated with the Standard Evaluation Protocol (SEP). The review results are submitted to the ministry.

Box 7.3. Research and education outputs of green education institutions

- The total number of students at Utrecht Faculty of Veterinary Medicine has been stable over time at around 1 500 each year. The number of Master's (or equivalent) diplomas has been around 200 per year.
- The total number of students at Wageningen University and the green HBO institutes has increased significantly, by 45% and 17% respectively over 2009-13. There are now about 8 300 students in Wageningen University and 100 000 in HBO institutes. As discussed in Section 5.3, this growth in enrolment is a huge contrast with the 1990s when Wageningen UR was formed. The reference projections (*referentieramingen*) show an expected stabilisation until 2030 for HBO, but further significant growth (50%) for Wageningen University.
- The growth is particularly high in the number of Master's students. Wageningen University has four times more Master's students than would be expected on the basis of the inflow of bachelors (primary data from the VSNU website).
- The total number of PhD graduates at Wageningen University has increased by 48% over the last decade (from 185 in 2000 to 273 in 2013), but not as significantly as elsewhere in the Dutch university system (75% increase) (VSNU and Wageningen UR, 2014).
- The quality of the courses and institutes of higher education is determined by the NVAO (Netherlands Flemish Accreditation Organisation). Wageningen University received its latest institute accreditation for a subsequent period of six years in 2012.
- The output of research activities has increased (as they have in other subjects and disciplines) over the last decades. Wageningen UR is considered to be one of the leading institutes in its field. As Spiertz and Knopff (2011) have demonstrated on the basis of Thomas Reuters data on publications in the field Wageningen UR is in the world top 5 concerning the number of publications and has an equally high number of citations per article. The QS world university ranking for agriculture and forestry puts the Wageningen UR on second place globally (2014).
- The quality of research activities is evaluated with the Standard Evaluation Protocol (SEP) which has recently been renewed for the next period from 2015 on. The most recent evaluations of Wageningen University research dates back from 2009.

Source: VSNU and Wageningen UR (2014), *Jaarverslag 2013*, http://www.wageningenur.nl/upload_mm/f/b/7/083f3a68-912c-4413-9794-e42e65533850_Jaarverslag%202013%20ENG_3.pdf.

Agriculture and food innovation and the general public

There is a good basis in trust in science by the general public, possibly linked to the good level of education, including in sciences (Section 5.3). In 2012 there was a survey on trust in science by the Scientific Council for government policy (WRR) and the Rathenau Institute. There was an expectation that trust in science was rather low due to scandals in science in the Netherlands. This, however, was not the outcome of a consultation of the general public.

Of all institutions in the survey (government, big companies, parliament, television, unions, newspapers, jurisdiction and science) science has the highest trust among the general public. This is confirmed by the Eurobarometer that gives the Netherlands a high score in trust in science. However, when scientists engage with politics/government or are commissioned by companies, trust drops to almost half.

In a public debate on trust in science (Rathenau/WRR, 11 March 2014) 'food and science' was one of the topics. The general idea was that although food science has brought valuable insights on the relationship between food and health, there is also much confusion about what to eat. The food scientists agree on one thing: whenever there is co-operation between science and business, always to be transparent about co-operation, especially when there is money involved. But even with transparency, trust remains an issue for the image of scientists working for or with businesses.

A related topic is trust in the safety of agricultural products. This trust has fallen due to a number of food scandals and negative media attentions. Whether or not this negative sentiment has influenced the attitude towards research institutes in the agricultural innovation system is not reported. A public debate has been going on about trust in food (13 January 2014 debate Schuttelaar en partners). 80% of the participants shared the general view that co-operation of food scientists and food industry is of vital importance.

Another related item is the trust in the food sector itself. This is first done in the first edition of the *Agrofoodmonitor* (Onwezen, 2013). The sector as a whole has a clearly positive image but that is not the case for all subsectors. The intensive livestock sectors (pigs and poultry) in particular have a less positive image.

There is little knowledge in the general public about how food is produced. Increased transparency should go along with communication on efforts to improve practices and product quality, reduce losses and increase recycling and animal welfare.

No specific reports have been found about trust in agriculture science or scientists.

An important role for the government would be to provide information on the benefits and risks of innovations, in a form accessible to a wider public, as is done through institutes like the Voedingscentrum.

Public and private investments in agricultural R&D

The public sector is the main source of funding for agriculture R&D, whether performed in public or private organisations. A wide variety of funding mechanisms are used from direct spending on research projects, including for Public-Private Partnerships (PPPs) and "pull mechanisms", to various forms of tax incentives. Business investment in R&D is normally driven by market demand, but governments also provide different kinds of incentives. Some, like R&D tax rebates, apply to the economy in general, while others are agriculture specific. In many countries, producer organisations and other non-governmental organisations also provide R&D funding (OECD, 2014a).

Macro-economic and budget statistics

According to macroeconomic data on national investment in R&D collected by the OECD, research intensity in the Netherlands, as measured by gross domestic expenditure on R&D (GERD) as a percentage of GDP, was comparable to the average of EU15 countries of about 1.8% over the period 2000-06. It then decreased slightly in the late 2000s, and increased after 2010 to reach 1.98% in 2013, which is slightly below the EU15 average of 2.07% (Figure 7.2). At 2.40%, the average share of GERD in GDP in OECD countries was slightly above the EU15 average due to higher rates in Korea, Japan, Switzerland and the United States (OECD, 2014c, Figure 3.4). Both EU and OECD indicators displayed a slight, regular growth over the period (Figure 7.2).

The GERD includes both public and private investments. For the Netherlands business expenditure on R&D (BERD) accounted for 1.14% of GDP in 2013 which is slightly more than half the total investment. BERD as a percentage of GDP in the Netherlands is below that of countries with advanced innovation systems such as Korea, Japan, Finland and Sweden but also lower than the EU15, EU28 and OECD averages (OECD, 2014c, Figure 3.6). On the other hand, with a rate of 0.63% of GDP in 2013, the Netherlands ranks among the top OECD countries in terms of the intensity of higher education expenditure on R&D (HERD) (OECD, 2014c, Figure 3.8). Another characteristic of R&D funding in the Netherlands is the relatively high share of GERD financed from abroad (over 10%) (OECD, 2014c).

Figure 7.2. Developments in R&D expenditure, 2000-13

Gross domestic expenditure on R&D (GERD) as a percentage of GDP

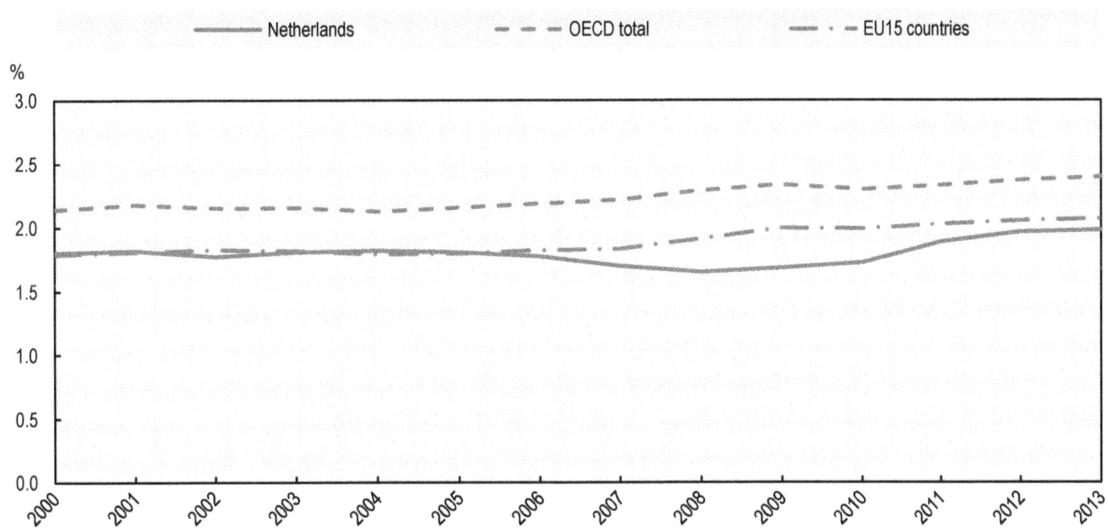

Source: OECD main science and technology indicators, OECD.stat. http://stats.oecd.org/Index.aspx?DataSetCode=MSTI_PUB.

StatLink http://dx.doi.org/10.1787/888933251175

The same information on R&D intensity and source of funding for the food and agriculture sector is not available at the sectoral level for the Netherlands in official statistics.[7] As far as business investments are concerned, OECD figures indicate that in the last year reported (2011) there was a total of EUR 563 million invested by businesses in R&D on agriculture, forestry and fisheries, and food and beverages, of which EUR 391 million comes from the food industry. This accounts for 0.79% on average of the total value-added of the sector; thus private investments in the agri-food sector lag behind those in the general economy.

If comprehensive information is not available at sectoral level, public budgetary efforts in R&D, as measured by Government Budget Appropriations or Outlays for R&D (GBAORD) can be compared for all activities and for agriculture as a percentage of corresponding GDPs. GBAORD intensity on agriculture was very similar to that for all activities until 2000, it grew faster in the early 2000s and in the mid-2000s, it was more than double. Agricultural R&D intensity then started to decline in the late 2000s and in 2013, it was 60% higher than that of all activities (Figure 7.3.A). Trends in GBOARD in constant prices also display the same variations over time (Figure 7.3.B). They show in particular that government budgets for agriculture have been much more variable over time than R&D budgets for all activities. At the end of a period of fluctuations, GBAORD for agriculture is about 7% lower in 2014 than it was in 1990. As a percentage of GDP, GBAORD for agriculture is much lower in the Netherlands than in Denmark, Germany, the United Kingdom or Norway, but it is higher than in France and Sweden (Figure 7.3.C).[8]

Figure 7.3. Developments in Government budget appropriations or outlays for R&D (GBAORD)

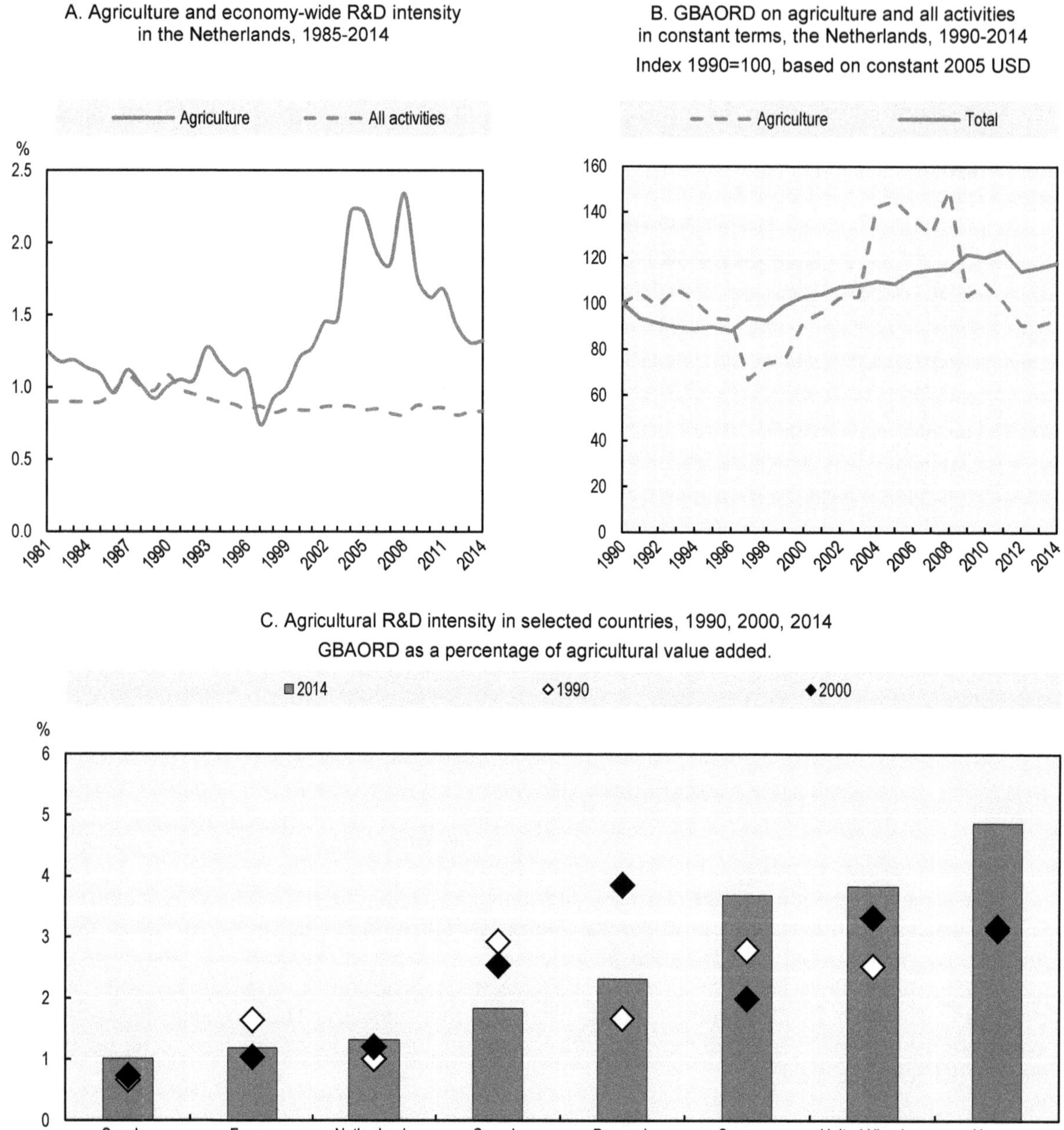

Data for Germany is 1991 instead of 1990. Data for United Kingdom is 2013 instead of 2014. Data for Canada is 2011 instead of 2014. For 2011, Canada national agricultural GVA is an adjusted aggregate of regional values. In 2006, Canada moved from ISIC REV 3 to REV 4.

Public expenditure on R&D is Government budget appropriations or outlays for R&D comes from OECD R&D Statistics, and value-added from OECD Gross Domestic Product statistics.

Source: OECD Research and Development Statistics, *OECD National Accounts and OECD Regional Accounts, 2015.* http://stats.oecd.org/.

StatLink http://dx.doi.org/10.1787/888933251187

Budget of agricultural institutes

The budgets for the primary actor of the agricultural innovation system, Wageningen UR (university and DLO) are another source of information on R&D investment on agriculture. The total annual budget of the DLO institutes has been more or less stable in the last decade with a budget of EUR 334 million in 2013 (Figure 7.4.A). Structural government funding is 43% of the DLO budget and tends to decline, while funds from contracts have a tendency to grow. These are all nominal budget figures, not corrected for inflation. At the moment a third of government budget for DLO is allocated to the top sectors Agri-Food and Horticulture and Propagation Materials. Government funding of statutory tasks accounts for another third, and government funding for research for evidence-based policy and knowledge base for the last third. In comparison, only 5% of government budget for TNOs is allocated to these two top sectors.

Within contracts those from companies have grown slightly from EUR 59 million in 2004 to EUR 61 million in 2013. The funding from contracts with the European Union have doubled, reaching a budget just over EUR 25 million in 2013 (Figure 7.4.B).

Figure 7.4. Turnover of DLO research institutes, 2004-13

Million EUR

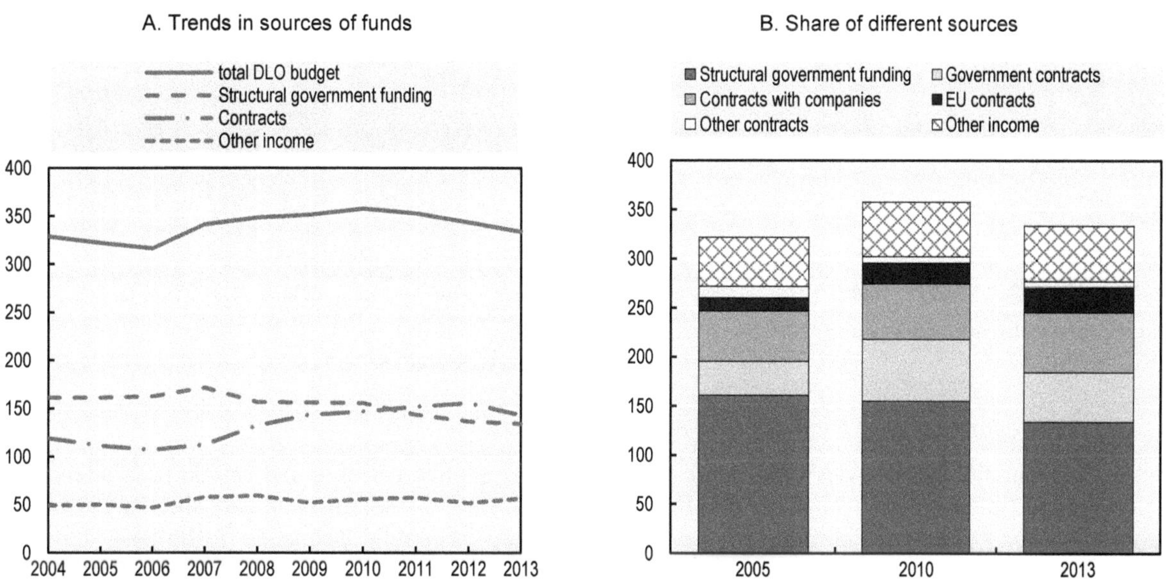

Source: Wageningen UR (2014), Jaarverslag 2013, http://www.wageningenur.nl/upload_mm/f/b/7/083f3a68-912c-4413-9794-e42e65533850_Jaarverslag%202013%20ENG_3.pdf.

StatLink http://dx.doi.org/10.1787/888933251192

Wageningen University shows a different picture over the last few years (Figure 7.5). The university is responsible for both higher education and fundamental research, and received structural funding from the government for those activities. The university also has contracts acquired on a competitive basis. Both structural government funding and assignments are growing, resulting in a 26% budget increase over 2007-13 (Figure 7.5.A). The growth in the number of students (bachelor, masters and Phds) is accountable for the growth in structural funding. The growth in contract funding reflects a doubling of the amount in contracts with companies (EUR 19 million in 2013) and an almost doubling of EU contracts (EUR 22 million in 2013) (Figure 7.5.B). The contracts are predominantly research contracts.

Government funding is a combination of education and fundamental research. There is no direct information on the share spent on research activities. Indications for the time use of academics at universities show that approximately half of total time is spent on research (Kok, de Jonge and Tom, 2007). What this means for the total budget of Wageningen UR for research is not specified.

The future for agricultural R&D government funding is uncertain. The figures produced annually by the Rathenau institute on total investment in research and innovation (TWIN) (van Steen, 2014) show a decline of public budgets for research in general as well as for DLO and TNO institutes. Based on a budget analysis of all ministries concerned there will be a budget cut of 14% over the period 2013-17 (calculation on the basis of TWIN 2014).[9]

Figure 7.5. Turnover of Wageningen University, 2007-13

Million EUR

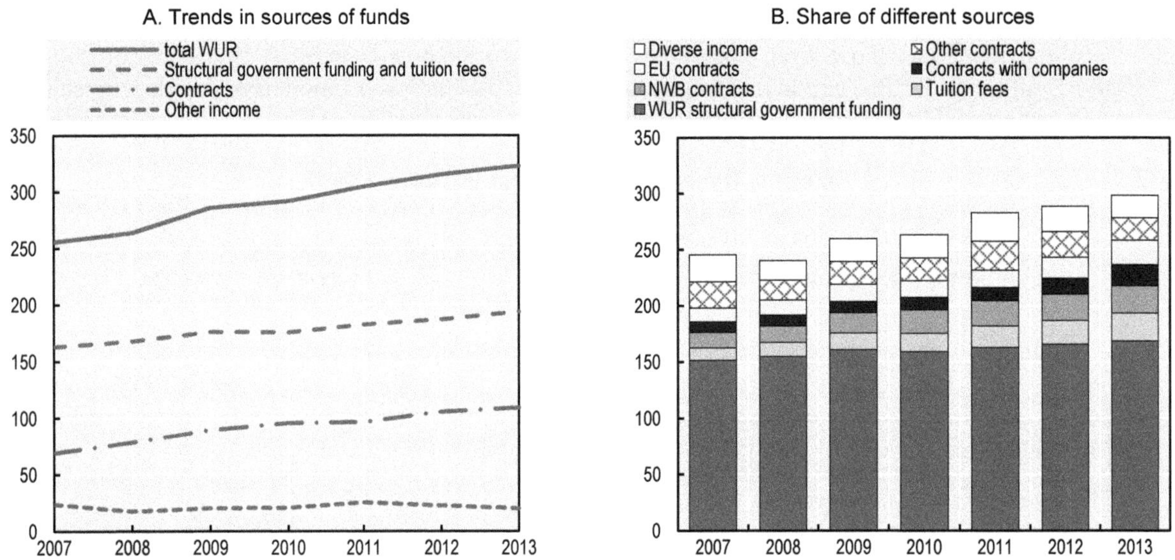

Source: Jaarverslagen Wageningen UR 2007-2013.

StatLink http://dx.doi.org/10.1787/888933251204

Private funding of public R&D and education institutions

Commodity board levies used to provide funding for applied research and certain programmes of vocational education. This source of funding disappeared when the boards were abolished in 2011. The next section notes the development of private funding of public academic institutions through competitive contracts, and in the context of PPPs, but the scope of activities is different, reflecting the interest of large firms, which have the capacity to engage in R&D.

Trends in funding mechanisms

Competition and structural funding

One of the trends in Wageningen UR budget is the growing amount of research funding in competition (Figure 7.5). There is a general tendency to spend more government funds in universities for specified subjects through principals of competition. Competitive mechanisms to allocate R&D funding have become more and more prominent and seem to dominate the academic landscape. Competition makes it possible to select the best proposals (and thus the best researchers) for a certain contract. Successful competitors achieve more money for doing research, are able to publish more (and in more prestigious journals). This helps them to generate more status and be more successful in subsequent competition. This trend is also visible in the agricultural innovation system.

Several mechanisms contribute to the growth of competitive funding in agricultural R&D:

- At the national level, there is an increasing flow of money through the Netherlands Organisation for Scientific Research (NWO) (Figure 7.6). The NWO promotes quality and innovation in science and falls under the responsibility of the Dutch Ministry of Education, Culture and Science. The NWO budget for DLO is not specified but for the Wageningen University the budget has risen from EUR 14 million in 2007 to EUR 24 million in 2013.
- Competition for contracts with companies – which are essentially competitive – is becoming more prominent. There is a growing budget for company contracts in the academic world in general and also in the agricultural innovation system. Again this is especially true and visible for the university within Wageningen UR. The amount of company contracts has doubled from EUR 9 million in 2007 to EUR 19 million in 2013. The DLO budget from companies is around EUR 60 million and growing slightly.
- EU contracts have also become a growing source of funding. This is not surprising as research budgets for the Horizon 2020 programme are significantly higher than those for its predecessor the seventh framework programme and EU funds are distributed through a competitive process. For both the university and the DLO institutes EU contracts have doubled in recent years to reach EUR 47 million in 2013 (of which EUR 22 million for DLOs and EUR 25 million for the university).
- The calls for the top sectors also are in competition. This will accentuate further the weight of competitive funding in the balance between competitive and structural funding.

Figure 7.6. Development of the NWO budget, 2001-12

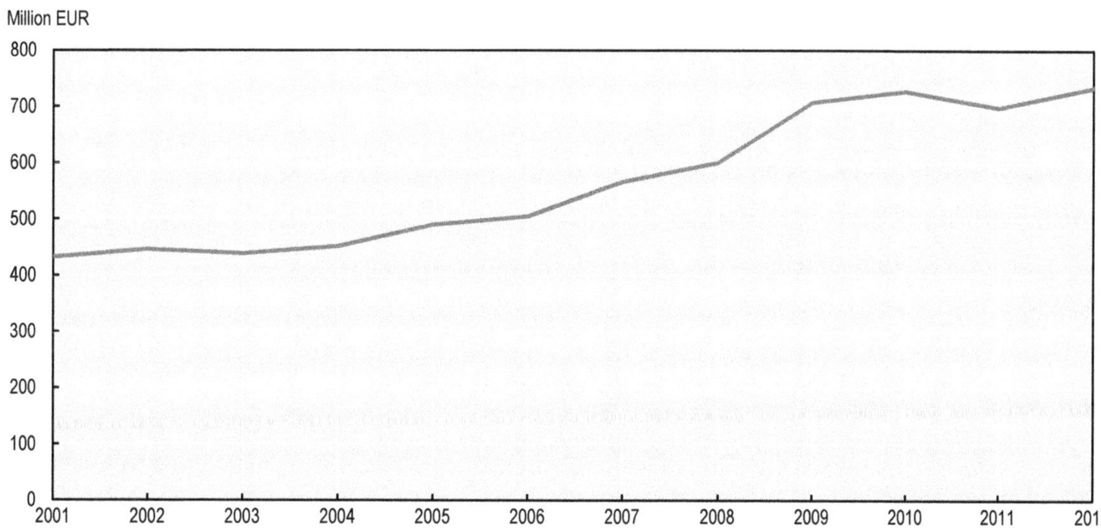

Source: NWO.

StatLink http://dx.doi.org/10.1787/888933251213

On the other hand structural funding for universities in general has decreased in the 2000s. The VSNU reported a decline from EUR 2 billion in 2000 to about 1 750 million in 2010. This phenomenon is not visible in the budget of Wageningen University because of the fast growing number of students in recent years (Figure 7.5). The decline in structural funding is, however, visible in former governmental institutes like the DLO institutes (Figure 7.4).

This tendency to decrease structural funding and increase competitive funding is in line with the principles of new public management to stimulate market functioning, competitiveness and efficiency of public services. However, competition comes to a certain price. A significant effort has to be made in the preparation of proposals in order for them to gain the first place. However, not all proposals succeed in getting funding and not all rejected proposals are insufficient ones. The scoring opportunities are lower than ever and it is very hard and even unlikely that even with excellent proposals one may get the contract.

Academics complain that a lot of the investments in proposals are eventually paid by the investments from the first flow funding. When the balance between the number of proposals (and the effort invested in them) and the gains (from awarded ones) is gone, competition itself comes with a high price.

A related price of competition is the discrepancy between the funds received from contracts and the integral costs incurred. A research activity by the WRR (scientific council for government policy) in 2004 and a specific case study on Wageningen University estimated that contracts cover 68% of the total costs of the projects involved (Schoutens, 2004). The remaining 32% is paid out of structural funding. According to a more recent investigation (Ernst & Young, 2014), the discrepancy between funds and costs is growing, both proportionally and in absolute terms. The discrepancy grows in absolute terms because project funding increased significantly over the last decade (and is expected to grow further in the coming years). The proportional discrepancy per Euro on contracts is also becoming larger: it represents EUR 0.74 per EUR 1 of project funding, which means that 57.5% of the real costs is covered by contracts, compared to 68% ten years earlier. And again the first flow funding is the source for this matching (and as such is falling).

Public private partnerships

The co-operation between firms and public knowledge institutes has for long been a widely acknowledged means of promoting innovation in the Netherlands. Spontaneously and with stimulation from local public authorities regional innovation hubs like Brainport Eindhoven, Seed valley Enkhuizen and Food valley Wageningen have evolved.

The top sector policy, described in Box 7.1, has been implemented since 2012 to activate public private partnership in research projects. Since October 2013 new innovation contracts have been signed for research activities covering 2014 and 2015. For 2014 the total investment covered by these contracts in all top sectors amounts to EUR 2.03 billion. These contacts cover all projects where co-operation between companies and public knowledge institutes is central. Almost half of the investment comes from companies (EUR 0.97 billion) and the rest (1.06 billion) is public investment.

The contracts for the two top sectors in food, agriculture and horticulture cover a total investment of EUR 249 million, of which about 30% (EUR 111 million) from the private sector. The DLOs expect to work with a budget of EUR 51 million for the top sectors in 2014.

It is too early to tell whether or not this strategy will be effective in the long run. In the short term there is a substantial amount of funds available for collaborative research. Panteia has developed a tool to monitor (among others) co-operation with knowledge institutes in the top sectors, which should facilitate the evaluation of progress in due time.[10]

Stimulating private investments through tax measures

As indicted earlier, there are three tax policy instruments to stimulate private investments. The overall budgets for these instruments have grown to EUR 756 million for the WBSO, EUR 302 million for the RDA and an estimated EUR 625 million for the innovation box in 2014. The total amount of tax reduction due to WBSO and RDA has for example increased by 25% between 2010 and 2014.

Primary agricultural production and agri-food industries benefit from WBSO and RDA tax reductions to a significant extent: EUR 65 million in 2013, of which EUR 31 million from WBSO and EUR 34 million from RDA. This amount has increased since the introduction of the WBSO in 1994, and following the introduction of the RDA in 2012, total tax reductions have increased by 50% compared to 2011.

The total number of companies concerned is rising gradually by 3.5% per year. Close to 800 firms benefitted from the WBSO, of which 75% also have RDA. They account for 1% of companies in primary production in the sector and 8% of companies in food industry.[11] The budget for the Innovation Box does not contain any sector-specific information.

The simplicity of the WBSO arrangement lies in the fact that it is solely oriented towards companies. So it is straightforward to implement. It also gives an indication of the total investment of business in R&D.

Comparative indicators

The significance of the agri-food sector for the Dutch economy is reflected in public investment in agricultural R&D (Table 7.1). Private investments on R&D, to a large part from the food industry, are slightly lower than the share of the food and agricultural sector in GDP, while public investments are in line.

Table 7.1. Comparative data on exports, private and public R&D investments for agriculture and the economy in general

		Netherlands total	Agri-food sector	Share of agro-food in total
		Million EUR	Million EUR	%
GNP[1]	2012	578 536	48 600	8.4
Export volume[1]	2013	426 200	79 200	18.6
Total investments in R&D (GERD)[2]	2013	12 728	1 016	8.0
Public investment in R&D[3]	2013	5 400	453	8.4
Private investment in R&D (BERD)[4]	2013	7 328	563	7.7
Tax reduction for R&D (WBSO/RDA)[5]	2012	6 450	261	4.0
GDP of top sectors[6]	2010	38% of total GDP	100% of agri-food GDP	23% of all top sectors
Investments top sectors	2014	2 030	249	12.3

1. Figures based on CBS data of Statline for the year 2012, and from Berkhout, P., H. Silvis and I. Terluin (2014), *Landbouw-Economisch Bericht 2014*, http://edepot.wur.nl/306953.

2. Total figures based on OECD Main science and Technology indicators (http://www.oecd.org/science/inno/msti.htm). Figures for the agri-food sector are based on the sources reported in this chapter .

3. Total figures based on OECD Main science and Technology indicators (http://www.oecd.org/science/inno/msti.htm). There is probably an underestimation for public investment in agricultural R&D because of the lacking figures of certain parts of the knowledge infrastructure like TNO food and RIVM nutrition.

4. Total figures based on OECD Main science and Technology indicators (http://www.oecd.org/science/inno/msti.htm).

5. Figures based on CBS data "gebruik instrumenten door topsectoren". www.cbs.nl/NR/...2534.../gebruikinstrumentendoortopsectorenmw.xls

6. CBS monitor topsectoren eerste meting 2010.

StatLink http://dx.doi.org/10.1787/888933251381

The other investments do not serve the food and agriculture sector as well. For the growing investment support through tax arrangements (WBSO and RDA) the companies in the two agriculture-related top sectors account only for 4% of all tax reductions.

Similarly, these companies do not seem to have access to investments in the top sector policy to the same extent as all companies on average. Agri-food companies account for 23% of the total value-added of the top sectors, as CBS reported for 2010 that the top sectors in account for 38% of the Dutch economy. In this perspective a share of investment of 12.3% of all top sector investments for agriculture-related sectors is much lower than one would expect on the basis of the significance for the Dutch economy and the relative position within the top sectors. This situation has not yet been acknowledged or explained.

General observations on trend in R&D funding

In addition to the difficulty to track policy efforts by sector, it is difficult in general to evaluate the extent to which the policies (fiscal arrangements, R&D expenditure and others) help to stimulate innovation. As outlined by Alston (2010) for example, it is difficult for theoretical and practical reasons to attribute an effect on agriculture to a specific expenditure or policy and to take account of long term effects.[12] According to the Court of Auditors' report (Algemene Rekenkamer, 2011), it is hard to evaluate whether or not all these innovation policies help in strengthening the innovative capacity of firms in the Netherlands.

The mix of innovative policies has been perceived by stakeholders as being complex, possibly due to the frequent policy changes and the growing share of competitive funding. Direct funding, project funding and competition for contracts, EU investment in Horizon 2020, the growing significance of tax measures for R&D activities by companies all add up to a very diverse landscape. Every actor tried to find its way in this landscape but for many it was too complex, as the Panteia study has shown (EIM, 2012). Co-ordination is of great importance within this context; the Court of Auditors has not been very positive on this. It should be noted, however, that the policy instrument landscape is now simpler than it has been for a long time.

Creating knowledge markets and networks

Intellectual property rights (IPRs), knowledge networks, and knowledge markets are of growing importance in fostering innovation, which increasingly requires collaboration and exchanges.

Intellectual property protection

Intellectual property protection is high in The Netherlands, as in many knowledge-based economies (Figure 7.7.C). Patent protection in particular has increased significantly in the 1980s and 1990s, to reach a level slightly below that of the United States, and following a similar pattern as other European countries (Figure 7.7.A). Plant Variety Protection as calculated by Campi and Nuvolari (2013) has also increased significantly to reach a level similar to that in the United Kingdom, lower than in Denmark and Germany, but higher than in France and the United States (Figure 7.7.B).

This has stimulated private investment in R&D in certain areas. Together with other incentives, it has contributed to the development of an export-oriented innovation sector. Strong IP protection also provides farmers access to foreign innovations. It should be noted, however, that not all innovations are or can be protected by IPRs, in particular non-technological innovations or in areas where fast adoption is required to maintain a competitive hedge, like horticulture.

IPRs are only used in specific area of research, like seed improvement, where they generate benefits well above their costs, but this is not often the case in horticulture or food processing, as mentioned by industry representatives.

Figure 7.7. Intellectual Property Protection

A. Patent Protection Index, 1960-2010
Score 1-5 (best)

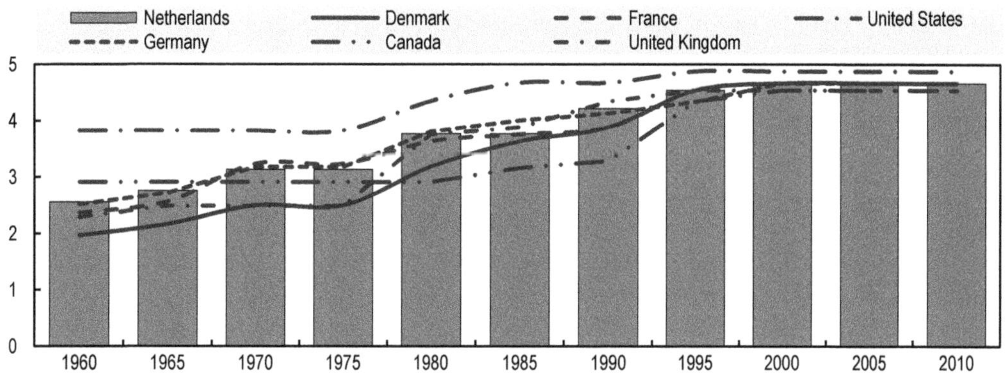

Sum of indices for duration, enforcement, loss of rights, membership and coverage.

Source: Unpublished update to the series from : Park, W. G. (2008), "International Patent Protection: 1960-2005", Research Policy, No. 37, 761-766. http://nw08.american.edu/~wgp/res_policy08.pdf

B. Plant Variety Protection Index
Score 1-5 (best)

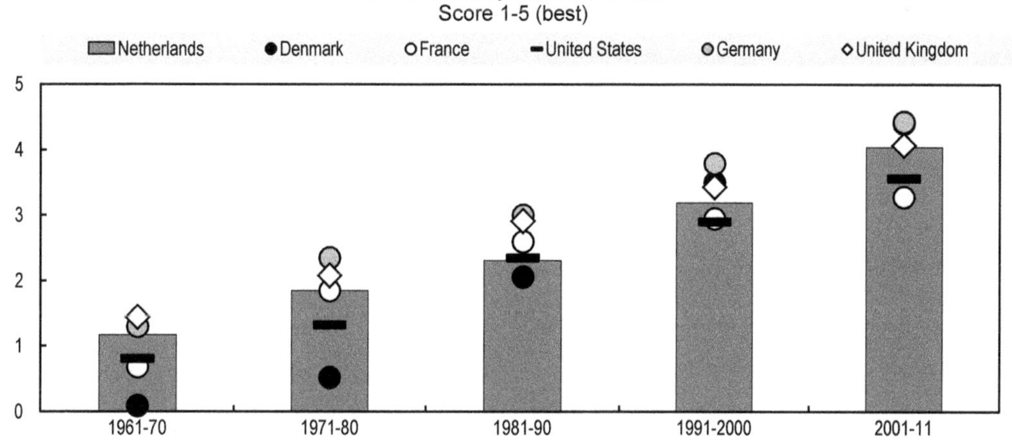

Source: Campi, Mercedes; Nuvolari, Alessandro (2013) : Intellectual property protection in plan varieties: A new worldwide index (1961-2011), LEM Working Paper Series, No. 2013/09 http://hdl.handle.net/10419/89567

C. WEF Intellectual Property Protection Index
Score 1-7 (best)

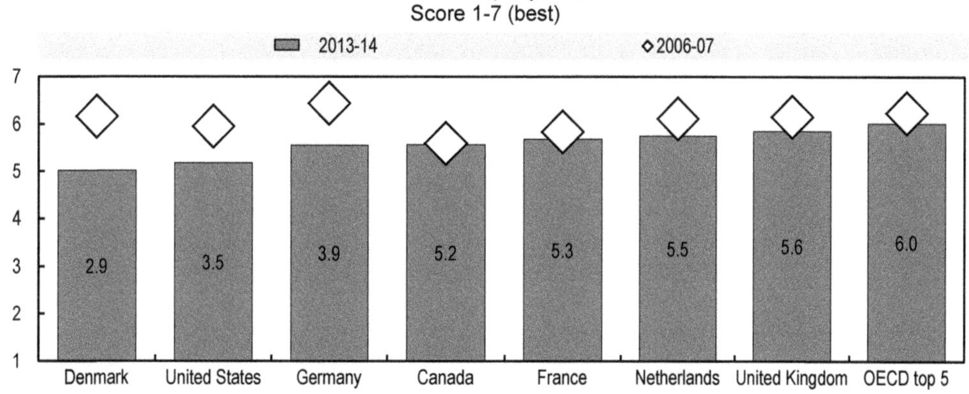

Source: World Economic Forum (2013), *The Global Competitiveness Report 2013-2014,* http://reports.weforum.org/the-global-competitiveness-report-2013-2014/#=.

StatLink http://dx.doi.org/10.1787/888933251226

IPR and plant breeding

The plant breeding sector is of high economic significance for the Netherlands with a steadily growing export value and a significant "spin off" to the trade in final products, in particular ornamentals. The Dutch plant breeding sector holds a strong position in vegetable crops, ornamental crops, and potatoes. The Netherlands plays a leading role in fundamental, strategic and applied research in plant genetics and plant breeding. The strong knowledge sector in the Netherlands is important for the plant breeding sector, including foreign seed companies that often have major R&D activities in the country.

Innovation in plant breeding is dependent on specific knowledge, the development and application of new technologies, access to genetic resources, and capital to utilise those factors. Access to technology as well as genetic material is essential for the development of new plant varieties. Competition and profitability of the plant breeding sector play a major role in the sustainability of the total food chain. Farmers and growers have an interest in competition in the seed market. Plant breeding is characterised by continuous innovations and the on-going development of new varieties that better meet the requirements of producers and consumers. The driving force behind this innovation is acquiring or increasing market share.

Technological developments showed a rapid progress in recent decades. One significant change results from the developments in molecular biology, initially outside agriculture, which led to the introduction of patent rights in the breeding sector. This system of intellectual property rights (IPR) certainly not only applies to genetic modification but to an ever broadening range of new techniques that make plant breeding more efficient and effective.

Patent positions in combination with technological developments have in recent decades led to large consolidation among breeding companies. For most crops only a few companies are controlling a large part of the world market. This makes a growing part of the global food supply dependent on a few companies. The access barrier for new companies to the plant breeding sector is high, where IPR plays a role next to the large amount of knowledge and expertise required to set up a breeding company and the long development period for new varieties. Dutch farmers and growers fear that their freedom of choice is threatened and that no varieties will be developed for certain crops that specifically meet their requirements when the decision power in breeding moves away from the Netherlands.

Based on Louwaars et al. (2009), Box 7.4 discusses plant breeders' rights and patent rights and their impact on the development of new varieties. The authors suggest amendments to patent rights that would facilitate further innovation.

The Netherlands is looking for support in the European Union for a comprehensive "Breeders' exemption". At the moment plant breeders have a limited breeding exemption, which was introduced in the patent law (Octrooiwet) in 2013 and took effect in July 2014.

A *limited breeding exemption* is an exemption under patent law which entails that anyone may make use of patented biological material for the purpose of breeding, or discovering and developing of other plant varieties without the consent of the patent holder. This limited breeding exemption does not apply to the commercial exploitation of the plant varieties thus obtained. When the patented feature is reflected in the new variety, the breeder will still have to obtain a license from the patent holder to trade this breed. In such negotiations, the breeder has a very weak position compared to the patent holder. Therefore, it should be possible for breeders to use plant varieties that are protected by patents. In the Netherlands, there is broad support for the introduction of a comprehensive breeders' exemption for patented varieties. A *comprehensive breeding exemption* is more extensive than the limited breeding exemption and could conceivably take any number of different forms, each with a greater or lesser scope.

A recent decision of the Enlarged Board of Appeal of the European Patent Office (March 2015) that confirms patentability of plants obtained by so-called "Essentially biological processes" such as crossing and selection, is a big setback for plant breeders. The decision may soon lead to hundreds of new patents on plant-related inventions.

Box 7.4. Plant breeder's rights and patent rights

The plant breeders' rights system is a specifically designed legal system for the protection of plant varieties. Plant breeders' rights give the developer of a new variety the right to exclude others from commercialisation. The breeder's exemption ensures that other breeders may in a sort of 'open innovation' use such a protected variety in their own breeding programme, making the best properties of these varieties available to the breeding programmes of competitors.

Plant breeders' rights and patent rights may be conflicting in plant breeding. Specific liberties of breeders and farmers are lost with the patentability of plant-related inventions. The significance of access to genetic resources for the development of new plant varieties was already recognised at the time of the Plant Breeders' Decree of The Netherlands ("*Kwekersbesluit*") in 1941 and has as 'breeder's exemption' been confirmed in more recent international treaties such as the International Union for the Protection of New Varieties of Plants (UPOV 1961/1978/1991), the Agreement on Trade-Related Aspects of Intellectual Property Rights (WTO-TRIPS - 1994), and the International Treaty on Plant Genetic Resources for Food and Agriculture (IT PGRFA - 2001).

Patent rights hold possibilities for strategic use, which may lead to lack of clarity in the market and to monopolistic behaviour. It may also lead to high costs of legal assistance. Plant breeders' rights have no such effects.

Recent analyses of the trends in genetic diversity of crops indicate that in Northwest Europe and North America genetic erosion has been brought to a halt and that diversity increases as result of a widespread use of genebank materials and new techniques, making use of such materials in breeding more effective. It is uncertain whether this trend is also visible at a global scale and whether it will continue when the number of breeding programmes diminishes as a result of further concentration in the sector.

The discussions about the roles of IPR in plant breeding also involve developing countries. These countries have difficulty in meeting the international IP protection requirements while at the same time optimising their IPR systems to meet the needs of their own society. Trade-related aspects of IP are dealt with within the framework of TRIPS under the Doha Development Agenda, focusing on the relationships between TRIPS and the Convention on Biological Diversity and the protection of traditional knowledge and folklore.

Access to genetic variation is essential for future crop breeding and IPR should stimulate innovative strength. The patent system would need to be amended to that end. This can be reached by: amendments of legislation and regulations, by increasing patent quality, and by improvement of the way that innovators use their patent rights.

The association of Dutch plant breeders suggests increasing the room for innovation in plant breeding by restricting the scope of patents in plant breeding, and more specifically by reinstating the exemption of patents on plant (varieties) or by introducing full breeder's exemption in patent rights. Both options would preferably be implemented at European level, possible via a revision of the Biotechnology Directive, and preferably in consultation with other countries with a significant plant breeding sector (such as the United States, Japan, and China). Implementation of such amendments may take a long time. Other policy options that could be introduced simultaneously are: tightening of the evaluation criteria for granting patents and banning the strategic use of IP rights that stimulate monopolistic tendencies in plant breeding.

Source: Louwaars, N. et al. (2009), *Breeding Business. The Future of Plant Breeding in the Light of Developments in Patent Rights and Plant Breeder's Rights*, Centre for Genetic Resources (CGN), Wageningen UR.

Knowledge exchange

The exchange of knowledge is facilitated in the Netherlands through the design of innovation policy, which requires collaboration between education, scientists in universities and research institutes, and businesses along the food chain down to retail and trade in the top sectors. In addition, some programmes dedicate specific support to participation in networks. The primary sector and small agri-food enterprises, however, face difficulties in being represented or participating in these activities as the role of the commodity boards in representing the commodity chain has not been taken up by other mechanisms.

It is not clear whether the organisation of R&D by sector facilitates or not knowledge exchange between sectors. Increasingly, the different top sectors exchange their experience. For example, the bio-based economy brings a lot of sectors together. However, knowledge circulation could be stronger between top sectors. In the agri-food sector, the specialisation of Wageningen University and the creation of Wageningen UR have facilitated exchanges within the sector, in particular between education and research and across agricultural disciplines, but also with the private sector. But collaboration on cross-cutting issues such as health and nutrition, or natural resource management, with researchers located in other universities, may be more difficult than in multi-purpose universities.

Over the year, centres of expertise or competitiveness, such as the food Valley, have been active but the reduction of public co-financing and its focus on top sectors threatens their vitality although the

Food Valley is carrying out work for the agri-food top sector. National funding has to some extent been replaced by provincial funds, which came from the privatisation of energy producing companies, but they will eventually run out. Provinces provide funds for SMEs through programmes (starts-up, thematic), grants and subsidies for innovation, and linking organisations, on a first-come, first-served basis.

R&D outcomes

Overall progress to create and adopt relevant innovations can be usefully monitored. Proxy measures, such as the number of patents of bibliographic citations, is available from international databases, including for primary agriculture and for upstream and downstream industries, and by type of innovation (OECD, 2014a).

The number of patents is not a comprehensive indicator of the outcomes of the innovation system, as not all innovations are patented, not all patents are used, other IPR systems exist for plant varieties, and trade secrets, rather than patents, are frequently used for food processing innovations. In addition, numbers should be complemented with indicators of patent quality, which are being developed at OECD (2013). This is, however, an informative proxy.

According to agricultural patent applications filed under the Patent Co-operation Treaty (PCT), which protects inventions in all signatory countries, the Netherlands is a significant contributor to the total of world agri-food patents covering specific agricultural inputs, agricultural sciences and food processing innovations, compared to its size, and the country's share in world agri-food patents is higher than the EU15 and OECD average (Table 7.2). The share of agri-food related patents in the country's total number of patents is higher than the EU15 and OECD averages but lower than the share of the corresponding sectors in GDP (Figure 2.1). As in other countries, most patents are in food processing rather than agricultural science.

Data on agricultural publications and citations confirm the significant contribution of Dutch agri-food R&D to innovation at the world level, which is higher than that of Belgium and Denmark, but lower than that of larger EU countries like France and Germany. The Netherlands' share of agricultural science publications and citations in the total world output has decreased in the last 15 years (Figure 7.8), reflecting the strong increase in contributions by BRIICS countries.[13]

Table 7.2. Agri-food R&D outcomes, 2007-12

	Netherlands	Belgium	Denmark	France	Germany	EU15 average	OECD average	OECD total
Agri-food specialisation: Agri-food science outputs as a share of country's total (%)								
Patents	8.8	11.7	11.3	5.3	4.4	6.9	5.6	..
Publications	6.9	6.8	10.2	7.1	6.4	8.4	9.4	..
Citations	6.4	7.7	8.7	8.0	16.9	10.8	11.9	..
Country's contribution to world agri-food science output (%)								
Patents	1.0	0.5	0.5	1.3	2.7	0.6	0.7	27.9
Publications	1.6	1.1	0.9	3.6	4.5	1.9	2.0	68.9
Citations	2.8	1.4	1.1	4.5	5.7	2.4	2.4	48.4

Source: OECD Patent Database, January 2014; SCImago. (2007). SJR — SCImago Journal & Country Rank. Retrieved March 19, 2014, from http://www.scimagojr.com.

StatLink http://dx.doi.org/10.1787/888933251390

Figure 7.8. Evolution of R&D outputs on agri-food sciences, 1996-2012

Percentage of world output

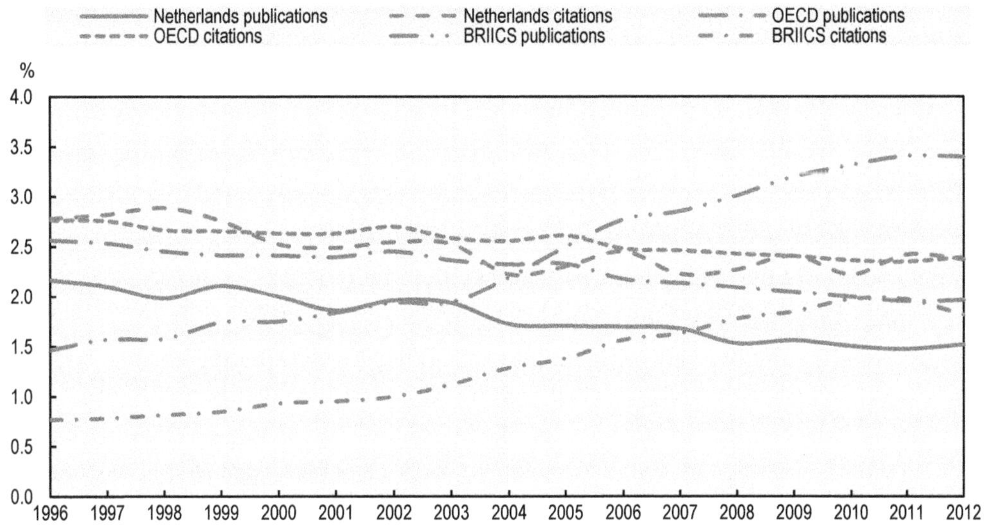

Agricultural sciences include Scopus journal classifications: agronomy and crop science, animal science and zoology, aquatic science, ecology/evolution/behaviour systematics, forestry, horticulture, insect science, plant science and soil science, and miscellaneous agriculture/biological sciences.

Source: SCImago. (2007). *SJR — SCImago Journal & Country Rank*, Retrieved 19 March 2014, from http://www.scimagojr.com.

StatLink http://dx.doi.org/10.1787/888933251237

According to a report by the Netherlands Environmental Assessment Agency (PBL, 2014), the Netherlands lags behind in terms of green (growth) innovation compared to neighbouring countries such as Denmark and Germany (its green-innovation index, based the share of green patents, is 80% of the OECD average and about half that of Denmark or Japan). Based on OECD patent, the Netherlands is, however, found to have a strong research position in agri-food biotechnology, non-biotechnology plant breeding and horticulture, and food chemistry (PBL, 2014, Figure 12).

Adoption of innovation in agriculture

Policy incentives for the adoption of innovation

The potential benefits of innovations are only realised if effectively implemented. Policy incentives for the adoption of innovation include a wide range of regulatory and financial approaches, including business investment support, and support to public-private co-operation arrangements and participation in networks. In primary agriculture, training, extension and advisory services can facilitate the transfer and successful adoption of innovation (OECD, 2014a).

There are numerous instruments to promote the adoption of innovations, which are used in the Netherlands:

- Introduction of new regulations that (gradually) enforce to change current practices, e.g. animal welfare regulations, or ammonia emission regulations.
- Public co-funding in public-private partnerships.
- Providing investment subsidies, including tax deductions.

- Providing guarantee facilities to adopters to spread the risks of adopters and banks, and help innovations cross the so-called "Valley of Death" or funding gap between basic research in the public sector and commercialisation of a new product by the private sector.
- Public (co-)funding of research, e.g. feasibility studies to demonstrate the (potential) value or to further develop the innovation; this may include vouchers that the government offers to SMEs to hire experts.
- Public (co-)funding of training, extension, innovation brokerage, and advisory services that can help adopters to select and implement innovations in a successful way; this may include vouchers that the government offers to SMEs to hire experts.
- Empowering entrepreneurs in a way that they become more skilled and confident to invest in innovative new products, services or even completely new concepts or business models.
- Public (co-)funding of network activities that bring together entrepreneurs, researchers and intermediates to learn from each other in order to co-create an innovation, rightly implement or further develop and innovation.

Table 7.3 shows the SME Innovation Stimulation Top sectors (so-called MIT) 2014 budget that the national government has allocated to the two agricultural top sectors of the Dutch economy in order to stimulate innovation by SMEs (OECD 2014c). It also shows how this budget was allocated between specific types of instruments to acquire knowledge about an innovation. The two top sectors received over a third of funds for R&D co-operation in SMEs, and the agri-food top sector alone close to a quarter of the total. Significant amounts were initially dedicated to feasibility studies and brokerage, in particular in horticulture.

Table 7.4 reveals, however, that the MIT 2014 budget allocation of Table 7.3 is a relatively small budget within the total budgets available for stimulating business innovation (OECD, 2014c). The TKI (Top consortia Knowledge and Innovation) allowance in 2013 totalled EUR 83 million, of which about 8% was allocated to "Agri-Food" and about 6% to "Horticulture and Propagation Materials".

Table 7.3. MIT 2014 budget allocation of the first tender by top sector and instrument

'000 EUR

Opening	15 April – 12 May				3 June – 22 September
Applicants	SMEs (FCFS)[1]		TKI (FCFS)[2]		SMEs (tender)
Instrument → Top sector ↓	Feasibility studies	Vouchers	Network activities	Innovation brokers	R&D co-operation
Horticulture and propagation materials	750		50	250	950
Agri-food	478.4				1 913.6
General budget (all top sectors)					8 000

MIT: *MKB-innovatiestimulering Topsectoren* (SME Innovation Stimulation Top sectors)

1. FCFS: First-Come, First-Served.

2. The TKI allowance scheme promotes the development of private R&D consortia in each top sector.

Source: RVO (2014a), "MKB-innovatiestimulering Topsectoren (MIT)", www.rvo.nl/subsidies-regelingen/mkb-innovatiestimulering-topsectoren-mit. (translated version). Reported in English in Table 5.7 in OECD (2014c), *OECD Reviews of Innovation Policy: Netherlands 2014*, http://dx.doi.org/10.1787/9789264213159-en.

StatLink http://dx.doi.org/10.1787/888933251405

Table 7.4. Business innovation funding instruments overview, 2014 or latest year

Main instrument	Annual budget (or annual average) EUR million	Additional public [source] or private funds leveraged EUR million	**Modality of delivery** (e.g. direct funding, fiscal incentives, loan or loan guarantees)	**Policy objective**	**Target population** (sector, size, age, innovative behaviour)
WBSO (Payroll tax allowance)	764	Private funds estimated[1] at around 1 390	Tax credit	R&D workers	All
RDA (R&D allowance)	302	n.a.	Tax credit	R&D, non labour costs	All
TKI allowance	102 (83 in 2013)	319 expected private funds (2013), 500 public funds aligned	Co-financing 25% supplement (40% for first EUR 20 000	Joint programming	Top Sectors only
MIT (SME innovation support Top Sectors)	30	n.a.	Choice of instruments[2] under discretion of top teams	SME participation in valorisation	Top Sectors only
MKB+ (SME+ Innovation Fund) consisting of:					
Innovation credit	86.5 (2013)	n.a.	Loan if successful, convert to a grant if project fails	Finance for high-risk innovation	SMEs, especially start-ups
Seed Capital	21.5	n.a.	Venture capital	Finance for innovation	High-technology entrepreneurs/SMEs
Fund of funds	100	50 from EU	Venture capital	Finance for innovation	High-growth innovative firms
Innovation box	625		Tax credit on profit from innovation	R&D investments	All
BMKB (SME Loan Guarantee Scheme)	705 (2012)	795	Loan guarantees (67.5% of loan up from 45% in 2013)	Facilitate credit, increased during the crisis	SMEs (14.6% in the top sectors receive 21% of budget)
NII (Netherlands Investment Institution)	[predecessor Syntens had 30.5 in 2013]	n.a.	Transfer of authority (agency)	Promote the availability of funding and facilitate investment	All
Microfinancing (by Qredits)	n.a.	30 (including private finance)	Micro-loan and business coaching	Finance for innovation	SMEs
Growth facility scheme (*Regeling Groeifaciliteit*)	50 (2012)	50	Loan guarantees (50% of loan)	Facilitate venture capital for SMEs	SMEs
Business Loan Guarantee Scheme (*Garantie Ondernemings-financiering, GO*)	329 (2012)	329	Loan guarantees (50% of loan)	Facilitate credit	Large and medium-sized firms
SBIR (Small Business Innovation Research Programme)	6.3 (2013) from central government	n.a.	Project funding for public procurement	Societal challenges, demand stimulation, valorisation of public knowledge	SMES, but partly open to large firms
Innovative Procurement Urgent (*Inkoop Innovatie Urgent*)	n.a.	n.a.	Project funding for public procurement	Societal challenges (demand stimulation)	n.a.

1. Estimated from public budget using the 1.77 "bang for the buck" estimate provided in EIM (2012), p. 12.

2. Choice of instruments includes collaborative R&D projects, feasibility studies, knowledge vouchers, hiring of experts, networking activities and innovation brokers.

Source: Ministry of Economic Affairs (2013), "Summary Chart Enterprise Policies" (unpublished), ERAWATCH website, Netherlands Enterprise Agency, RVO (2014b), WSBO/RDA programme, 15 April 2014 and correspondence with Ministry of Economic Affairs. Reported in English in OECD (2014c), *OECD Reviews of Innovation Policy: Netherlands 2014*, http://dx.doi.org/10.1787/9789264213159-en, Table 5.8.

Further, Table 7.4 reveals that by far the largest part of the innovation support to firms consists of tax credits and loan guarantees, while venture capital or seed money for developing an innovation is scarce. This is detrimental to agriculture, which consists of many micro-(family)-enterprises with few or none employees, no R&D department and typically low profit margins (and consequently little room for tax deduction). In this situation, adoption of innovation in the sector would gain from stronger government involvement in defining priorities and from more targeted instruments (Van der Vlist and Van Galen, 2005).

The three rows in Table 7.5 show how the abovementioned instruments are directed towards policy tasks to enlarge the pool of innovators or to better exploit the available capacity. The columns show how the instruments stimulate different types of innovation with respect to their newness in a specific context.

Instruments that support the adoption of innovation are particularly important for small and medium-sized enterprises (SMEs), which are predominantly present in the agriculture and food complex, because these farms and firms do not have an R&D department and no or few innovation specialists among their employees. Moreover, it is often more difficult for SMEs to get loans from banks for innovations because the risks are typically high relative to the size of the farm or firm as a whole. However, access to public funds for innovation can be difficult for them because of the multiple sources, the complexity of rules and delays in delivery. The horticulture and food processing industry is made of many SME+ companies, which are too large to be eligible for support for very small businesses, but often too small to hire specialists in fund raising. Horticulture companies hire tax specialists to benefit from tax incentives. They find it easier than to prepare a project to obtain direct support. They find public money so costly to obtain that they do a cost-benefit analysis to assess whether it is worth asking. Efforts are being made to address this issue, the Netherlands Enterprise Agency (RVO) and a single desk platform launched in 2014 (*ondernemersplein*) provide information regarding programmes and regulations and advice to entrepreneurs for free.

Table 7.5. Business support policy mix according to current capacity and further development needs

Capacity building / development stage → ↓Policy tasks↓	A. […from no innovation activity to innovation that is] *new-to-the-firm*	B. […from primarily new- to-the-firm to innovation that is] *new-to-the-market*	C. […from new-to-the-firm and new-to-the-market to innovation that is] *new-to-the-world*
1. Increase the pool of innovators	WBSO, RDA, micro-financing	WBSO, RDA; MKB+, BMKB (small firms); TKI, MIT, RDA+ (Top Sectors only)	WBSO, RDA (only partially due to small scale); Innovation Box; MKB+, BMKB (small firms only); TKI, RDA+ (top sectors only)
2. Increase the intensity of innovative effort	WBSO, RDA, Growth Facility; MKB+, BMKB	WBSO, RDA; MKB+, BMKB (small firms); TKI, MIT, RDA+ (Top Sectors only); Business Loan Guarantee Scheme	WBSO, RDA (only partially due to small scale); Innovation Box; MKB+, BMKB (small firms only); TKI, RDA+ (top sectors only)
3. Diversify by extending the range of innovation modes and fostering collaboration	WBSO, RDA	TKI, MIT; RDA+ (Top Sectors only); SBIR (very partially due to small scale)	TKI, MIT; RDA+ (Top Sectors only); SBIR (very partially due to small scale)

See list of programmes in Table 7.4.

Source: OECD (2014c), *OECD Reviews of Innovation Policy: Netherlands 2014*, http://dx.doi.org/10.1787/9789264213159-en , Table 5.9.

Innovation in practice

The education and skills policies and the agricultural innovation system provide the essential underlying elements for the adoption of innovation, namely: the development of new knowledge, insights and technologies, the availability of researchers, teachers, intermediates, entrepreneurs, and skilled employees who can be part of the co-creation process in which the new knowledge, insights and technologies are converted into innovative products and services and implemented in practice. A high level of general and scientific education may also facilitate acceptance of technological innovation by society at large (OECD, 2014a).

Access to training and advisory services

Farmers in the Netherlands are generally highly educated and have the capacity to look for information on the innovations enabling them to remain competitive. As seen in Chapter 7, there is a diversity of institutions providing advice to farmers on a wide range of topics. Knowledge transfer is a competitive market and there is no evidence of inadequate supply or insufficient access to those services. Moreover, farmers' access to training, extension, innovation brokerage, and advisory services is supported using EU pillar 2 measures. In RDP 2007-14, Measure 111 for vocational training, information actions channelled about EUR 5.8 million per year of EU and national funds to Dutch farmers (PSE database). Grants are awarded for courses or training sessions that are not part of regular agricultural education programmes.

Developments in the advisory system raise, however, some concerns among experts (Verstegen et al., 2013). With the privatisation of the public advisory service DLV, farmers that could collect the necessary information themselves, or had no problem hiring advisors, continued to innovate. Others fell behind and had to stop. The consequences of privatisation for the advisory organisations are also large. Farmers hiring (and paying) advisors are more critical on getting value for money and more and more request very specialised advice, e.g. on legal issues or farm expansion. This has given room to new (specialised) entrants in the advisory business, leading to increased competition, lower profit margins, and therefore more difficulties for advisors to stay up-to-date. It is expected that, as in various other branches, many advisors will be solo entrepreneurs (so-called ZZP'ers), which will raise the competition. Increasingly specialist advisors participate in network projects and applied research projects allowing them to stay up-to-date and keep contact with the farmers.

As discussed in Section 5.3, agricultural education is offering adult training courses covering technical and managerial knowledge, as well as entrepreneurship programmes that help valorisation of knowledge. These programmes are defined in co-operation with the industry.

Farm innovators and followers

About 14% of enterprises in the Dutch primary agricultural sector made a significant improvement in the products or processes in 2012. The proportion of "real innovators" – companies that first introduced a new product or production process – is estimated at less than 2% of all firms (Figure 7.9). These estimates are reported by LEI, which gathers data about the innovative behavior of entrepreneurs in the agricultural and horticultural sector in its annual survey (Van der Meer, 2014).

The shares of innovative firms vary in time and per sector. For example, poultry production was the most innovative sector of agriculture and horticulture in 2011. As in 2010, there was much renewed in the poultry sector because of the welfare requirements for buildings. Particularly in horticulture, a continuous decrease in the number of innovative enterprises has been observed in recent years. The poor operating results undoubtedly play a role. In greenhouses especially, floriculture firms implement product innovations (new species/varieties). Process innovations in greenhouses are made by both floriculture and greenhouse vegetable companies. Frequent renewals are: sorting and packing lines, climate control and lighting. For arable farms, the introduction of GPS broadens. The use of GPS has several advantages, in the form of higher yields, more efficient application of fertilizer and mechanical weed control (Van Galen, 2012, 2013).

Figure 7.9. Development in innovation diffusion in Dutch farms, 2005-12

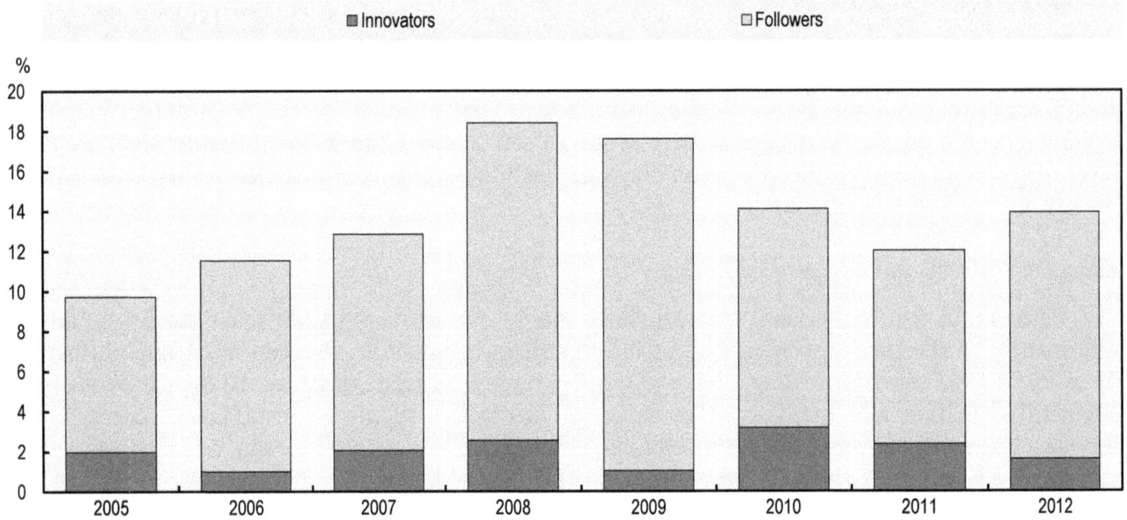

Source: Van der Meer R.W. (2014), "Innovatie in de land- en tuinbouw 2013", http://edepot.wur.nl/317866.

StatLink http://dx.doi.org/10.1787/888933251247

Innovation in food processing

Information on innovation adoption in food processing firms is available from the Community Innovation Surveys published by Eurostat. Figure 7.10 illustrates the range of innovation activities performed in food and drink companies, from R&D and training to acquisition of machinery and marketing of innovation in 2006-08.[14] The main activities reported by Dutch companies (about half of them) are domestic R&D and acquisition of machineries, equipment and software. In comparison, three-quarters of French companies report domestic R&D activities. Dutch companies are less involved in training than French and German ones, but more engaged in R&D abroad. Figure 7.11 sheds light on the extent to which food and drink companies collaborate in product and process innovation with other companies or organisations. Companies collaborate mainly with suppliers and private R&D institutes. Dutch companies collaborate slightly more than French ones but less than Danish ones.

Obstacles for innovation

Since Rogers (1962) it is widely acknowledged that besides education and skills policy many other aspects play a role in the adoption of innovations including attributes related to the innovation itself, e.g. compatibility, complexity, visibility, relative advantage, and testability. Another issue limiting the adoption of some innovations is public acceptance of certain types of innovations, e.g. Genetically-Modified Organisms and innovations requiring large-scale farms to be operational and profitable such as biogas operations.

A study on adoption of innovations in greenhouse horticulture revealed that the influence of the firm advisor on the final adoption decision should not be underestimated. Of 39% of the horticultural firms, the probability that advisors take the final adoption decision is higher than 50% (Verstegen et al., 2003). Other aspects that play a role are uncertainty in government policy, the overall economic climate, the economic perspectives in agro-food, the solvency and cash flow of farms and firms in agribusiness, the family-firm life cycle, the availability of a farm successor, the credit policy of banks, the risks involved with the required investments, and the extent to which innovators and early adopters can benefit, e.g. from innovation subsidies, first-mover advantages, or intellectual property rights, before imitation takes place (Van Galen and Verstegen, 2008).

Figure 7.10. Share of food and drink enterprises engaged in innovation activities, by type of activity

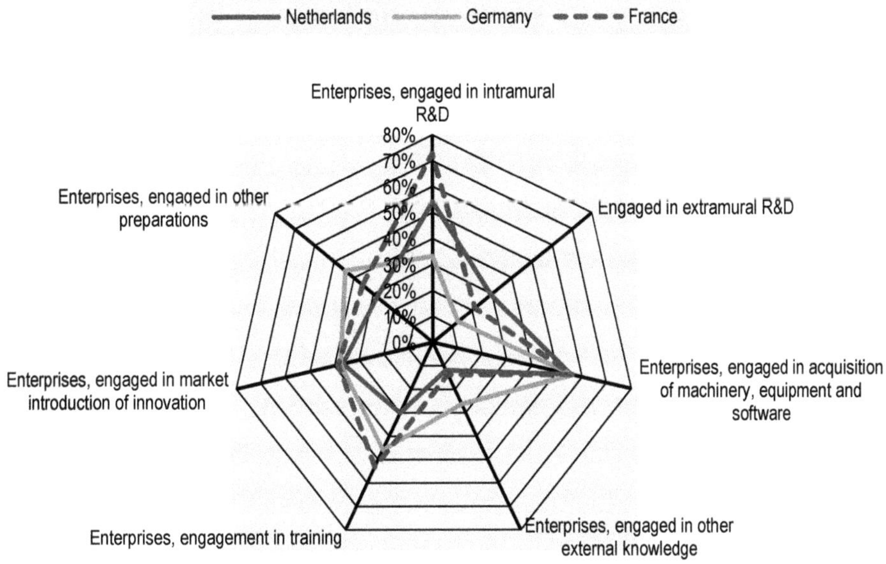

Source: Eurostat CIS, 2006-2008, calculations LEI. In: Van Galen, M. et al. (2013), *Jukema, Innovatie in de levensmiddelenindustrie; Een internationale benchmarkstudie*, LEI Wageningen UR.

StatLink http://dx.doi.org/10.1787/888933251257

Figure 7.11. Share of enterprises that collaborate in product and process innovation with other companies or organisations, by origin

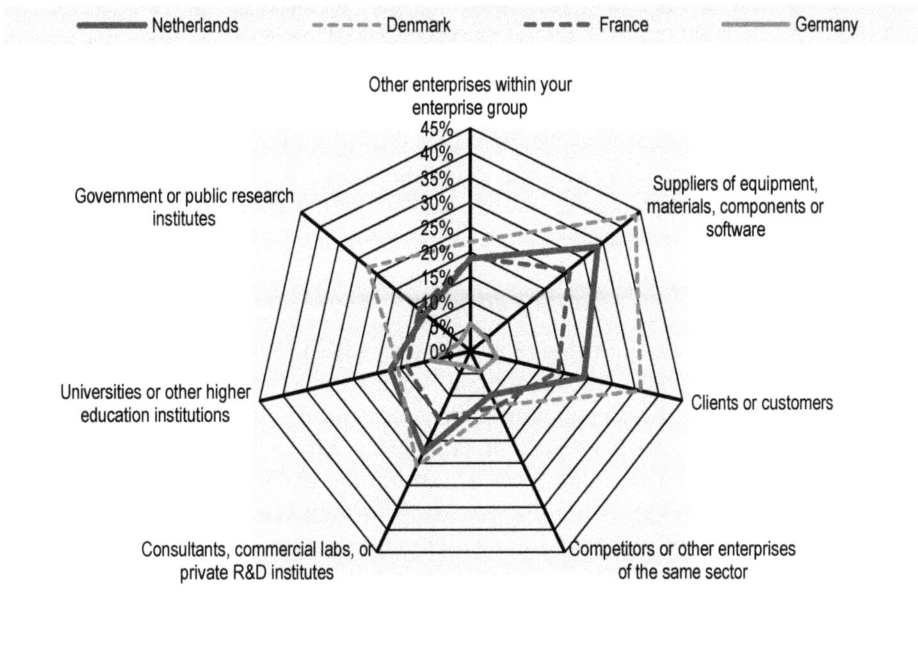

Source: Eurostat CIS, 2006-2008, calculations LEI. In: Van Galen et al. (2013), *Jukema, Innovatie in de levensmiddelenindustrie; Een internationale benchmarkstudie*, LEI Wageningen UR.

StatLink http://dx.doi.org/10.1787/888933251267

Figure 7.12. Obstacles to innovation in agricultural firms

Percentage of agricultural firms indicating that a specific obstacle was important or very important for their innovation decision

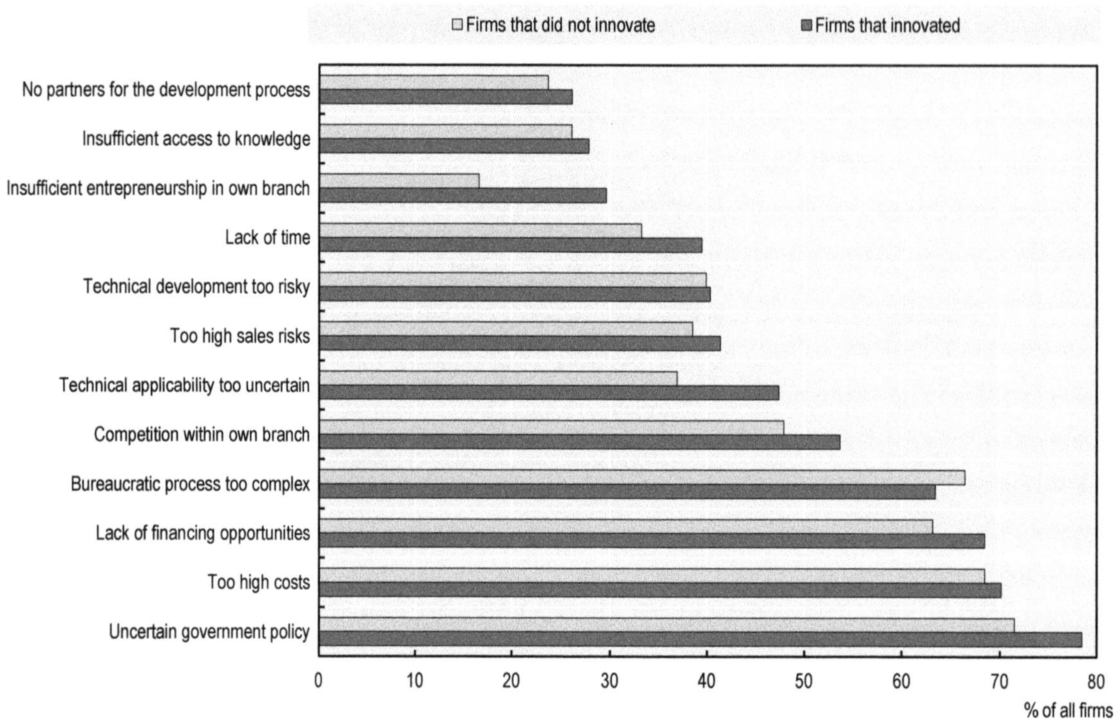

Source: Van der Meer, R.W., and M.A. Van Galen (2013), "Innovatie in de land- en tuinbouw 2012." http://edepot.wur.nl/286031.

StatLink http://dx.doi.org/10.1787/888933251279

Over 60% of the farmers and horticultural growers in the Innovation Monitor of the Dutch Farm Accountancy Data Network indicate that uncertain government policy, too high costs, lack of financing possibilities and too complex bureaucratic processes are important obstacles for innovation (Figure 7.12). The complicated financial toolbox in particular is favourable to the expansion of existing firms but not for new firms and new ideas.

Reflections on recent developments regarding adoption

Increasing competition in the knowledge system may lead to more value for money for farmers. Yet, a number of factors lead to suboptimal investments in innovation development and implementation. In addition to financing issues, inherent to any investment, there are also specific failures in knowledge markets, such as lack of knowledge capacity, high costs and risks for early adopters, low property protection on some types of innovation, or lack of adapted solutions due to the diversity and fragmentation of agricultural systems, which makes it difficult to identify demand. Government intervention is therefore crucial to facilitate the adoption of innovations needed to improve productivity growth, sustainably, when markets fail to do it, by increasing access to knowledge and financing, guiding R&D investments in areas that are not profitable in the short term for the private sector, and fostering the participation of farmers and SMEs in agricultural systems.

With its many micro (family) enterprises agriculture benefits from a government that puts forward clear policy goals and instruments, facilitates in forming innovation agendas to set (R&D) priorities and fosters the collaboration among firms, and between firms and knowledge institutions. Until a few years ago, the Dutch Ministry of Agriculture not only could facilitate the formation of innovation agendas but could also directly execute large parts of these agendas by the various Green knowledge institutions.

With the start of the top sectors policy, agribusiness firms have obtained more influence on the concrete research activities at these institutions. Especially when they co-finance research projects they tend to guide the research project towards their own firm development issues, rather than towards collective issues and research questions in Dutch agriculture. And generally the research may focus more on less risky topics with short-term-benefits. Moreover because the commodity boards have been abolished, important revenue streams for the applied research institutes have disappeared causing a large reduction in the number of collectively-financed research projects. The consequence of all this is that it will become more difficult for micro-enterprises in agriculture (without their own R&D facilities) to derive innovative technologies and services, feasibility studies and so on from the collective knowledge institutions.

It is difficult to predict what the effects will be on the adoption of innovations. Because the education level of agricultural entrepreneurs has increased considerably in the last decades, one could argue that agricultural producers could develop innovations themselves, or co-create these in interaction with peers and/or other supply chain partners. However, not every entrepreneur will have the skills to compensate for the fact that innovations are no longer presented in an accessible way. As with many developments, this development will also speed up the process in which firms are separated in firms that can deal with the new context and firms that fall behind and eventually have to close. This will reduce the pool of farmers that potentially can come up with innovative ideas. At the same time with ICT developments the world has become a village and that farmers have the opportunity to find inspiration across the globe.

International co-operation in agricultural R&D

Context

International co-operation in agricultural research and development (R&D) offers universal benefits because of the public good nature of many innovations in agriculture. The benefits of international co-operation for national systems stem from the specialisation it allows and from international spill-overs (OECD, 2014).

As the general innovation system, R&D on food, agriculture and horticulture is highly internationalised, with a strong participation in EU programmes, international networks, and programmes developed at national level to promote international co-operation. In addition, Wageningen University is open to foreign students and staff.

European co-operation in agricultural R&D

International partnership is a common R&D activity for Dutch knowledge institutes. European programmes such as the Framework Programmes for Research and Technological Development, Interreg, Knowledge and Innovation Communities (KICs), Joint Programming Initiatives (JPIs) and the ERA-Nets are common ground and there is broad experience in many programmes, at different levels of coordination and participation.

The **Seventh Framework Programme for Research and Technological Development (FP7)** was the EU's Research and Innovation funding programme for 2007-13. FP7 funded different programmes grouped under different headings according to their main purpose: capacities, co-operation, ideas, people, and co-ordination of research activities.[15] The specific programme on co-operation, which supports all types of research activities carried out by different research bodies in trans-national co-operation, received the bulk of the funding. ERA-nets and Joint Programming initiatives (JPI) were funded under the heading co-ordination of research activities. Food, agriculture and biotechnology was originally allocated a budget of EUR 1.9 billion over 2007-13, of which EUR 1.76 million were spent by the end of 2013.[16]

The current programme for funding EU research and innovation is called "Horizon 2020".[17] Nearly EUR 80 billion of funding will be available over seven years (2014-20), compared to EUR 55.8 billion over the previous period 2007-13.[18] Over the period 2014-20, EUR 3.8 billion have been budgeted for "Food security, sustainable agriculture, marine and maritime research and the bioeconomy", representing a doubling of budget funds compared to the previous programming period.

Participation of agriculture-related top sectors in FP7

To compare FP7 participation within the top sectors, there are three important parameters: return rate (share of available FP7 budget obtained by Dutch participants), success rate (share of proposed projects that have obtained financing) and the amount of financing obtained. In Figure 7.13 these parameters are related: the more a top sector is positioned in the upper right corner, the better the performance. In that case, both a high success rate and a high return have been achieved. The lines in the graph indicate the average return and success rates of Dutch participants. The area of the circle is a measure of the funds obtained by Dutch organisations in the respective top sector. The top sectors "Agri-Food" and "Horticulture and Propagation Materials" are not the biggest in terms of funding, but they are performing well: they have a high success rate and in the highest return rate of all top sectors.

Figure 7.13. Success rate, return rate and amount of finance obtained by Dutch participants]

Million EUR

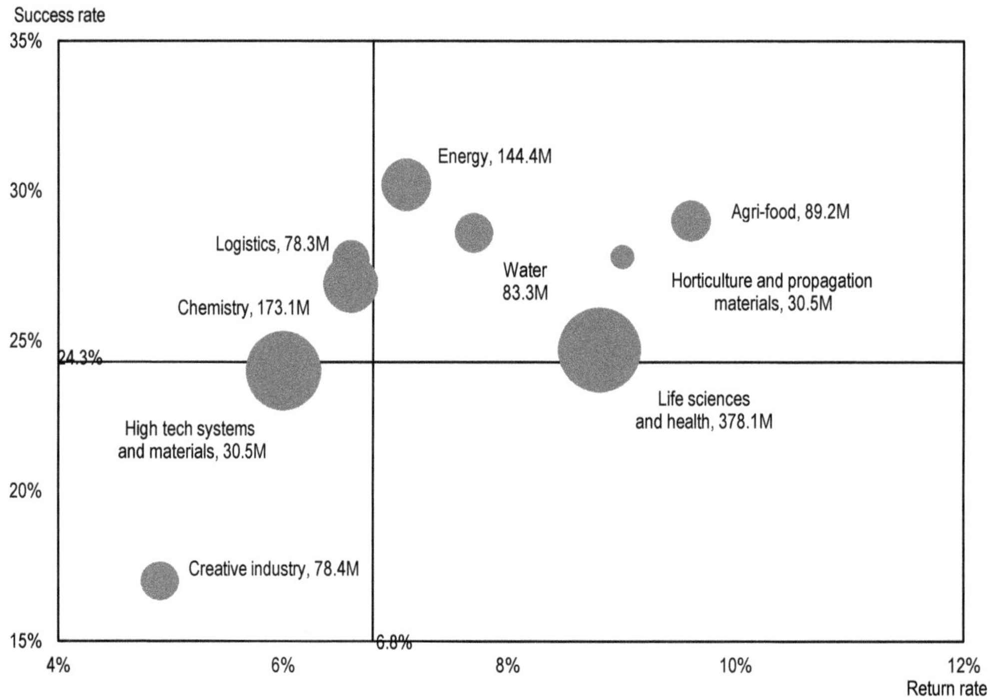

Source: Agentschap NL (2012), *Nederlandse topsectoren in KP7 2011*, Ministry of Economic Affairs, Agriculture and Innovation. https://view.publitas.com/agentschapnl/nederlandse-topsectoren-in-kp7-2011/#page/1.

StatLink http://dx.doi.org/10.1787/888933251288

Top sector Agri-Food

Dutch knowledge institutions and businesses are active participants in FP7 co-operation projects on agri-food. They were awarded EUR 89.2 million in financial support (Table 7.6). EUR 10.8 million, or 12% of this amount, were allocated to Dutch businesses, of which two-thirds to SMEs, including non-commercial organisations. Close to 10% of the total EU budget available for agri-food in FP7 co-operation went to the Netherlands, a high rate compared to a share of 6.8% of overall FP7 budget and compared to the size of the Dutch sector at EU level (about 6% of EU total value of agricultural production in 2014).

With one or more organisations the Netherlands is represented in 64.4% of a total of 275 projects selected. Compared to the average share of the Netherlands in FP7 Co-operation (41%) Dutch organisations are thus well represented in EU Agri-food projects. In comparison with other countries in the top 10 of the top sector Agri-Food, the Netherlands achieves the highest success rate (29%). The Netherlands receives about as much funding for projects in Agri-Food as France and Germany, which are much larger countries. Only the United Kingdom receives a substantial larger amount. Italy and Spain are also major players in this field.

Agri-food projects are financed within the Co-operation theme Food, Agriculture and Fisheries, and Biotechnology (KBBE). Important research areas are: Increased sustainability of all production systems, Nutrition and Food Processing. The Netherlands achieves a high proportion (88.9%) and a high return (15.4%) for projects in Nutrition. The results in Food quality and safety are above average. To Dutch organisations a high amount of funding is allocated for co-ordinating projects in this area.

The following organisations are the top six in the Agri-Food top sector based on the number of closed FP7 contracts over the period 2007-10:

- Foundation for Agricultural Research, part of Wageningen UR (70 contracts).
- Wageningen University, part of Wageningen UR (34 contracts).
- VU University Amsterdam and VU University Medical Centre (7 contracts).
- National Institute for Public Health and Environment (7 contracts).
- Radboud University (5 contracts).
- Ministry of Economic Affairs (5 contracts).

In the top sector Agri-Food most frequent collaborations are with organisations from the United Kingdom (298 times), Germany (248 times), France (234 times), Italy (200 times) and Spain (182 times) (Agentschap NL, 2012).

Table 7.6. Dutch top sectors Agri-&Food and Horticulture and Propagation Materials in FP7 (2007-10)

	Unit	Agri-food	Horticulture and propagation materials
Total FP7-funding:	Million EUR	930.5	341.1
Funding to Dutch partners:	Million EUR	89.2	30.5
- Higher or secondary education	Million EUR	35.2	8.3
- Research organisations	Million EUR	41.3	19.0
- Large business	Million EUR	3.9	0.8
- SMEs	Million EUR	6.9	2.5
Dutch matching funding	Million EUR	33.5	10.4
Total number of allocated FP7 projects		275	96
Number of projects awarded with Dutch participants		177	60
Share of projects with Dutch participants	%	64.4	62.5
Share of EU funding to Dutch participants	%	9.6	9.0
Success rate of projects with Dutch participants	%	29.0	27.8
Overall FP7 success rate	%	16.3	16.3

Source: Agentschap NL (2012), *Nederlandse topsectoren in KP7 2011*, Ministry of Economic Affairs, Agriculture and Innovation. https://view.publitas.com/agentschapnl/nederlandse-topsectoren-in-kp7-2011/#page/1.

StatLink http://dx.doi.org/10.1787/888933251411

Top sector Horticulture and Propagation Materials in FP7

Dutch knowledge institutions and companies are assigned EUR 30.5 million in financial support from FP7 participation in projects in the field of horticulture and propagation materials. Eleven percent of this amount is allocated to Dutch businesses, with Dutch SMEs, including non-commercial organisations, receiving most of it.

Dutch participants received 9.0% of the total available budget, compared to close to 10% for Agri-Food and 6.8% overall. The Netherlands with one or more organisations were participating in 62.5% of the 96 projects selected under this topic. Compared to the average participation of the Netherlands in FP7 Co-operation (41%) Dutch organisations are also well represented here.

The Netherlands achieves the highest success rate (28%). Organisations from the United Kingdom and France received the largest amounts in this field. The Netherlands receives an amount approximately equal to that of Germany. Italy and Spain receive less financial support. Projects in the field of horticulture and propagation materials are also funded within the Co-operation theme Food, Agriculture and Fisheries, and Biotechnology. The main research area is: Increased sustainability of all production systems.

The following organisations are in the top 5 in the Horticulture and Propagation Materials top sector based on the number of FP7 contracts signed over the period 2007-10:

- Foundation for Agricultural Research, part of Wageningen UR (26 contracts).
- Wageningen University, part of Wageningen UR (13 contracts).
- Keygene (3 contracts).
- Leiden University (3 contracts).
- Radboud University (2 contracts).

In this sector, collaboration is mainly with organisations from the United Kingdom (97 times), France (89 times), Germany (75 times), Spain (74 times) and Italy (69 times).

Standing Committee on Agricultural Research (SCAR)

The Standing Committee on Agricultural Research (SCAR) includes representatives of EU member states, with representatives from candidate and associated countries as observers. In total 37 countries participates to the SCAR. It is presided over by a representative of the European Commission, and has a mandate to advise the Commission and the member states on the coordination of agricultural research in Europe.[19]

The SCAR committee was given in 2005 a renewed mandate by the EU Council to play a major role in the co-ordination of agricultural research efforts across the European Research Area (EU SCAR, 2013). The SCAR committee adopted a structured approach to the prioritisation of research topics for further collaboration, through the establishment of a number of Collaborative Working Groups (CWGs) and Strategic Working Groups (SWGs) formed by Member & Associated State representatives.

The establishment of CWGs is a more flexible and less formal alternative mechanism to the ERA-NET scheme funded under FP7 collaboration, but shares the same objective: to stimulate and ultimately increase research collaboration between funders and programme managers on key-research areas. In some cases, the term "Strategic" has been chosen to characterise those CWGs which have no vocation to become ERA-NETs and are more long-term policy driven.

Since 2005, more than 20 CWGs/SWGs have been set up by European countries engaging voluntarily and on a variable-geometry basis in the definition, development and implementation of common research agendas based on a common vision of how to address major challenges in the field of agricultural research. The dynamism and commitment of several CWGs paved the way for participation in the FP7 ERA-NET scheme. The initial discussions in SCAR CWG were taken on by ERA-NETs.

ERA-Nets for agri-food research

ERA-Nets in which agri-food and environmental cross-border issues are central seem to work well, particularly for animal welfare and diseases, plant health, food, seafood, organic farming, ICT/robotics and integrated pest management) (EU SCAR, 2013).

Cross-border co-operation allows collaboration with others to tackle problems within the same geographical/climate area, decreases fear of collaboration at transnational level (breaking language and cultural barriers) and provides training for future projects at the EU level. It promotes the exchange of experiences from different regions and countries with the same production specificities while creating networks for future actions. Moreover, cross-border collaboration is important where a clear mutual interest is identified in complying with EU regulations (pesticides, nitrates, soils) or emerging/spreading threats (animal, plant diseases, invasive species, climate change).

Examples of ERA-NET actions selected during FP7 (2007-13) with participation of the Netherlands

- EMIDA (Coordination of European Research on Emerging and Major Infectious Diseases of Livestock): Project Ref: 219235.
- ICT-AGRI (Coordination of European Research within ICT and Robotics in Agriculture and related Environmental Issues). Project Ref: 235460.
- RURAGRI (Facing Sustainability: New Relationships between Rural Areas and Agriculture in Europe): Project Ref: 235175.

In the Work Programme 2014/15 of Horizon 2020, the successor of FP7, more than 20 ERA-NET Cofund actions and a number of Coordination and Support Actions (CSAs), including those in support to Joint Programming Initiatives (JPIs), are planned. The budget of the EU dedicated to ERA-NET Cofund actions amounts to close to EUR 93 million in 2014 and more than EUR 163 million in 2015.

Global co-operation

CGIAR

The CGIAR is a global partnership that unites organisations engaged in research dedicated to reducing rural poverty, increasing food security, improving human health and nutrition, and ensuring sustainable management of natural resources.[20] It is carried out by 15 Centres, which are members of the CGIAR Consortium, in close collaboration with hundreds of partners, including national and regional research institutes, civil society organisations, academia, development organisations and the private sector.

The CGIAR Consortium develops, integrates and coordinates CGIAR Research Programmes across the 15 Centres and partners. The Consortium Board has ten members, including an *ex* officio member, the Chief Executive Officer of the CGIAR Consortium. One of the ten members of the Board is prof. Martin Kropff, vice-chair of Wageningen University and Research centre. He has been appointed as the Director General of the International Maize and Wheat Improvement Center (CIMMYT), starting 1 June 2015.

The Netherlands government supports the CGIAR financially. In 2012, it committed more than EUR 165 million to the CGIAR Fund over the four year period 2012-15.

Global alliances on agricultural research

The Netherlands participates in a number of recent global alliances on agricultural research, such as the Global Alliance for Climate-Smart Agriculture, launched on 23 September 2014 at the UN Climate Summit by the President of Niger and the Prime Minister of the Netherlands, to improve people's food security and nutrition in the face of climate change.[21] The Netherlands is also a founding member of the Global Research Alliance on Agricultural Greenhouse Gases launched in 2011.[22] Members of Wageningen UR co-chair the Livestock Research and the Group Inventories and Monitoring Cross-Cutting Group.

The Netherlands is also a member of the Collaborative Research Network on Sustainable Temperate Agriculture (TempAg), which has been launched at OECD on 23 April 2015. TempAg will assist Members to respond to emerging challenges such as "sustainable intensification" (driving agricultural productivity gains while using inputs more efficiently, and lowering pressure on the environment by reducing losses of water and nutrients, and greenhouse gases from agricultural lands), and "resilience" (the development of farming systems and landscapes that respond positively to economic and climatic shocks).

International co-operation in higher education

Nuffic is the Netherlands organisation for international co-operation in higher education. As an independent, non-profit organisation based in The Hague, the Netherlands, Nuffic supports internationalisation in higher education, research and professional education in the Netherlands and abroad, and helps improve access to higher education worldwide.

International scope of Wageningen UR

Research collaboration

In 2013, Wageningen UR participated in projects in more than 110 countries worldwide, ranging from small projects (with a budget between EUR 10 000 and EUR 100 000) to projects involving millions of euros. There were two regional Wageningen UR offices at the end of 2013: In Chile (Latin America) and in China. Various Wageningen UR Sciences Group have offices in countries such as Bangladesh, Saudi Arabia, Uganda and Zambia.

Generally, international funding has increased. Wageningen UR receives a large share of some EU programme funds, which indicates that Wageningen scientists have strong networks within Europe. An increasing part of the research funding is provided by other funding agencies, such as the European Research Council and at the global level the World Bank, and the Melinda and Bill Gates Foundation.

Participation in international networks contributes to the international orientation of the agricultural research system in the Netherlands. Many alumni of Wageningen University have the opportunity to acquire a short-term or even permanent position abroad, either as scientist in the United States, in CGIAR institutions or as an assistant-expert for the Food and Agriculture Organization (FAO) to support developing countries.

Wageningen UR seeks an expansion of alliances with leading international universities and research institutes, beside the renovation of existing memorandums of understanding (MoUs) with international partners and the continuation of activities in the Euroleague for Life Sciences (ELLS) network (Box 7.5). Details of the activities in third countries (United States and Canada, Latin America, Asia, Africa) are presented in the Wageningen Annual Report 2013 (Wageningen UR, 2013).

Box 7.5. Global mandate at Wageningen UR

In her address "Bridging the gap between sciences and society" at the opening of the academic year 2014/15, Professor Louise O. Fresco (2014) spoke about the global mandate of Wageningen UR:

"We are asked to respond to grand societal questions – as the European research programme Horizon2020 calls them so eloquently. We all agree about the overall challenge for the future: the transition to a circular bio-based economy where all outputs, including waste and emissions, become inputs in new production processes, where ecosystems are protected and restored in order to contribute to well fed, healthy and sustainable societies and individuals. Notwithstanding lots of technical progress (e.g. recyclable biomaterials) the details of this overall transition are far from clear, and multiple trade-offs exist, between biofuel and food production and between different bio-based sources. Designing new technology alone is insufficient without social change. In some food chains two thirds of the greenhouse gas emissions occur at the household level, e.g. through cooking or refrigerating. More efficient resource use may lead to more consumption, unless consumers adapt their behaviour. So the challenge of the future is not just technological but also one of values and behaviour.

There remains one enormous issue: the gap with the bottom billion. Clearly, there cannot be sustainable development with rampant poverty, inequality and civil strife. The poor need to be connected to markets, to health care and education. We know that economic growth is a necessary, if not a sufficient condition to reduce poverty. Most of the poorest countries have sizeable rural populations: there economic development can only mean agricultural growth. Tackling unemployment through value-added entrepreneurship in food chains is the way to provide a decent living to youngsters. More of them need to be trained as scientists too, in Africa in particular. The UN [United Nations] Sustainable Development Goals provide a new and hopefully useful framework which we are well placed to contribute to.

Our mandate is global, our science knows no borders. Yet we are still predominantly a Dutch organisation, in administration and culture, even if we teach mainly in English, and are proud of our foreign students and staff. Undeniably Dutch pragmatism has been part of our success. However, we need a greater diversity of minds and cultures and scientific views. An institution where the light shines in all colours. To attract and retain the greatest talents in the world we must make them feel at home, scientifically and personally.

Like food production, Wageningen UR has local roots and global reach. There can be no doubt about our international scope. The world needs our knowledge and our graduates. We will build even stronger networks, in developed and developing countries. We also want to join forces with our European, North American and Pacific colleagues. Some of the world's best science and most innovative companies are located there. Of course, working in a global environment entails risks of overstretching in countries where private and public sector governance may be deficient, so a cautious ethical stance is called for."

Source: https://www.wageningenur.nl/en/newsarticle/Fresco-Bridging-the-gap-between-sciences-and-society.htm.

International students

The education provided by Wageningen University is multidisciplinary and very internationally focused, both in the curriculum and the composition of the student population. Around 45% of the students come from abroad.

As of the reference date of 1 October 2013, non-Dutch Bachelor's students at Wageningen University accounted for 6% of the total number of undergraduates (247 out of a total of 4 231 students). Approximately 80% of these students are from Germany. Of the total number of Master's students, approximately 40% are non-Dutch students (1 658 of 4 190), spread across 96 nationalities. In total, close to half of international students came from European countries (Table 7.7). The main countries of origin of foreign Master's students are China (391), Greece (161), Germany (128), Indonesia (75), Ethiopia (66) and Spain (64). All Master of Science programmes are in English and are provided for Dutch and non-Dutch students alike.

Table 7.7. International Bachelor's and Master's students at Wageningen University (excluding exchange students)

Nationality	Reference date 1 October 2013
European Economic Area countries	898
Europe (excl. European Economic Area)	21
Africa	179
Asia	646
North and South America	155
Other	6
Total	1 905

Source: Wageningen UR.

StatLink http://dx.doi.org/10.1787/888933251422

Erasmus Mundus is a valuable programme for Wageningen University to give further content to international partnerships and joint Master's and PhD programmes. In 2013, there were three joint Master's programmes, two joint PhD programmes and two partnerships with third countries, focusing on the mobility of students and staff. In 2013, a proposal for a new partnership was approved by the European Union. Within the scope of one of the joint PhD programmes, the first PhD degrees are expected in 2014. One of the joint Master's programmes has reached the end of its funding period. This programme will continue under the Erasmus Mundus Brand Name.

International Exchange Students

The number of outgoing students of Wageningen University increased from 235 in 2013 to 323 in 2014. The number of incoming students was 357 in 2012/13 and 397 in 2013/14.

Human resources

Wageningen UR has given special attention to support staff from abroad who enter employment at Wageningen UR (Table 7.8). Frameworks have been developed for sending staff on longer-term stays abroad. This has resulted in more transparency for employees, management and human resources. Wageningen UR's Travel Policy states that if a travel warning applies to a certain country or human rights are being violated there, that country should not be travelled to unless a manager approves this with higher management or the unit director.

Table 7.8. Scientific staff of Wageningen University, 31 December 2012 (in fte) towards nationality and HOOP-region

Country/region	Full-time equivalent	Share
Netherlands	1 072	68.2%
Germany	63	4.0%
Belgium	18	1.1%
Southern Europe	84	5.3%
Eastern Europe	57	3.6%
Rest of Europe	62	3.9%
Other western countries	37	2.4%
Turkey	7	0.5%
Africa	20	1.2%
South and Central America	39	2.5%
China	52	3.3%
India	25	1.6%
Other Asia	26	1.7%
Middle East	12	0.8%
Total	1 573	100.0%

Source: www.vsnu.nl/nieuws/nieuwsbericht/153-nederlandse-onderwijs-en-onderzoek-steeds-internationaler.html.

StatLink http://dx.doi.org/10.1787/888933251433

Measurement of R&D collaboration

About 27% of Dutch agriculture patents have a foreign co-inventor (Table 7.9), indicating that Dutch agricultural innovators are more engaged in cross-country cooperation than the average of EU28 and OECD countries. As for other countries, collaboration rates are higher for publications than for patents. At about 65%, the collaboration rate is higher than the OECD and EU15 averages. It is close to those in Belgium and Denmark and higher than in France or Germany. While government and EU policy influence collaboration, it is also strongly linked to the size of the country, with innovators in smaller countries more likely to collaborate internationally when they have a certain level of research capacity.

Table 7.9. Agri-food R&D co-operation, 2006-11

Agri-food outputs with co-authors as a share of total agri-food outputs (%)

	Netherlands	Belgium	Denmark	France	Germany	EU15 average	OECD average
Patents	27.1	49.8	22.1	36.6	30.9	17.0[1]	11.8
Publications	65.1	63.0	64.3	58.8	55.2	57.7	50.8

1. EU28.

Source: OECD Patent Database, January 2014; SCImago. (2007). SJR — SCImago Journal & Country Rank. Retrieved 19 March 2014, from www.scimagojr.com.

StatLink http://dx.doi.org/10.1787/888933251442

Reflections on recent developments in international R&D co-operation

As strong performers in agri-food R&D, Dutch research and education institutions are active and successful partners in collaborative efforts on food and agricultural research, in particular at the EU level, but also in global networks and initiatives. As a result, the share of EU funding in the budget of R&D institutions increases over time. Excellence in agri-food research and education attracts foreign investment from foreign multinationals interested in research collaboration. Wageningen University is a highly ranked institution at the world level, which attract foreign staff and students. This has positive short and longer term benefits for the international orientation of the system.

There are, however, concerns about the capacity of the system to maintain its contribution to collaborative efforts. On the one hand, the stability of the Dutch system becomes increasingly dependent on EU funding, with the reduction of government expenditure on R&D. On the other hand, competition for obtaining foreign research funding becomes more intensive and transactions costs increase. Moreover, collaboration requires national co-funding. But national funding is decreasing in general and in particular in areas not covered by the top sectors. In some areas of less interest for businesses but with strong public good components (such as policy-relevant activities), Dutch researchers struggle to find co-financing for EU projects. Finally, international collaboration requires the maintenance of world-level capacities over time.

Summary

- The Dutch agricultural innovation system is a very good performer at the national and international levels. The good performance of the agricultural innovation system came from strong investment over the long term, and tripartite collaboration between education, research and industry. Both have led to high R&D supply, accumulation of knowledge stock, and good knowledge infrastructure, allowing for both international collaboration, and the development of solutions adapted to the sector's demand.
- Dutch research and education institutions grouped under Wageningen UR are strong performers and active and successful partners in collaborative efforts on food and agricultural research, in particular at the EU level, but also in global networks and initiatives.

- Adoption of innovation in farms and firms is widespread. Well-educated farmers have access to a diversity of training and advisory services on a wide range of technical, organisational, management and marketing aspects, which have facilitated the adoption of innovation. The agricultural universities of applied sciences and the agricultural colleges contribute to lifelong learning and innovation.
- Institutional developments have made the agricultural innovation system more collaborative and demand-driven and have strengthened the role of the private sector in guiding investment. R&D co-ordination and funding mechanisms have facilitated industry-driven projects, public-private partnerships and networking, at the sub-sector and regional levels.
- Government expenditure on agricultural R&D has declined in recent years, increasing dependence on foreign funding, in particular from the European Union, and private funding.
- A high and increasing share of funds is project-based and delivered through competitive mechanisms. While this trend increases the relevance of R&D activities with regard to objectives of funders, it leads to higher transaction costs, uncertainties and difficulties in funding risky projects.
- Another development is the shift of responsibility for the knowledge base and knowledge infrastructure to shared public-private responsibility.
- Tax incentives provide the majority of support to innovation in the private sector. There have also been changes in instruments used to support innovation in the private sector, with the role of tax incentives increasing while direct support declined. This helps private research institutions directly and public research institutions indirectly because it encourages the industry's participation in Public-Private Partnerships (PPPs).
- The top sector policy introduced in 2011 concentrates public funds for innovation on nine export-leader sectors. The top sector policy subjects the granting of public funding for applied research in R&D institutes to participation in public-private partnerships (PPPs) within top sectors.
- Two top sectors are dedicated respectively to the export orientated "agri-food" sector and "horticulture and propagation materials" sector. They receive a healthy share of all top sector public funds. However, the top sector contribution to the financing of public research institutions is limited by the effectively lower share of private co-funding and the fact that the private contribution is often in kind or financed through tax incentives. As a result, the top sector policy cannot fully compensate for the decline in public funds in the long term.
- The top sector approach strengthens collaboration between actors and ensures R&D activities are geared towards innovations that will improve the economic performance of the top sectors. However, it carries some pitfalls for the long-term: business-driven innovation tends to focus on low risk and short term R&D activities, and generally invests less in fundamental and public good-related research, even though some large companies may have a longer term approach. The ambitions of the private sector and the government are changing rapidly, making it difficult for long-term R&D to adjust, and for knowledge institutions to pursue long-term objectives. There are also concerns about the capacity of the system to maintain its contribution to international or regional collaborative efforts, given trends in funding levels and mechanisms.

Notes

1. The term "Agricultural Knowledge and Innovation System (AKIS)" is used in the European Union to describe the features of systems producing agricultural innovation, with the same meaning as the term "Agricultural Innovation System" (AIS) used in the OECD and the World Bank for example.

2. Since 2010 TNO is under the governance of the Minister of Economic Affairs.

3. The Centre for BioSystems Genomics is a consortium of major Dutch and international companies and top plant scientists working on potato, tomato, Arabidopsis and Brassica. It is a unique public-private partnership in plant genomics: http://www.cbsg.nl/.

4. In 1998 public advisory services were privatised to form DLV Advisory Group Ltd (DLV *Adviesgroep NV*), which included many specialised groups. After a reorganisation around 2006-07, four groups continued as separate, independent entities: *DLV Plant*, which specialises in crop an horticulture (http://www.dlvplant.nl/uk/content/over-dlv-plant.htm); and *DLV Dier*, which specialises in livestock husbandry (http://www.dlvdier.nl/dlvconsultancy) and includes former DLV Rundvee Advies BV, DLV Intensief Advies BV, DLV Bouw, Milieu & Techniek BV, and DLV Makelaardij BV, as well as Aequator Groen & Ruimte BV, which specialises in environmental and rural development (http://www.aequator.nl/welcome-website-aequator-groen-ruimte-bv/), and DLV Agriconsult BV.

5. The OECD Territorial Review of the Netherlands recommends aligning the Top Sector innovation strategy with the EU regional cluster policy in order to provide more coherent local incentives (OECD, 2014d).

6. The WBSO is described in the website of the Netherlands Enterprise Agency as: 'WBSO is an Act that provides a fiscal facility for companies, knowledge centres and self-employed persons who perform R&D work'. http://english.rvo.nl/subsidies-programmes/wbso-rd-tax-credit-and-rda-research-and-development-allowance.

7. Table 8.1 present, however, unofficial estimates of private investment in agri-food R&D reconstructed using available information.

8. For more information on GBAORD trends in OECD countries, see OECD (2013), Table B.2.

9. Twin figures on total investment in research and innovation are available at: http://www.rathenau.nl/publicaties/publicatie/voorpublicatie-totale-investeringen-in-wetenschap-en-innovatie-twin-2013-2019.html.

10. http://www.panteia.eu/About-Panteia.

11. These figures come from the most recent source: a RDO report 2013 (Focus). Using a different classification, CBS estimates R&D tax reduction at EUR 71 million for 2011 and EUR 92 million for 2012. The CBS uses the classification of the topsectors as their basis for collecting figures on WBSO and RDA. The topsectors Agri-Food and Horticulture and Genetic Materials together pretty much cover the whole food, agriculture and horticulture sector in the Netherlands. The report of RDO uses another classification of firms. RDO specify agriculture forestry and fishery as a whole and the food industry as a separate category. These two classifications do not seem to match.

12. Alston (2010) outlines data and methodological issues related to the evaluation of the economic impact of public expenditures on agricultural R&D and reviews estimates found in the literature.

13. Brazil, Russian Federation, India, Indonesia, the People's Republic of China, South Africa.

14. The more recent Community Innovation Survey covers 2012. The data set is described in http://ec.europa.eu/eurostat/web/microdata/community_innovation_survey, and results for all

firms surveyed are available at: http://ec.europa.eu/eurostat/documents/2995521/6483064/9-21012015-BP-EN.pdf/ad7e4bf6-fc8f-459b-a47e-da1c9043bf2e. Access to micro-level data is needed to obtain results for food and drink companies.

15. FP7 website: http://cordis.europa.eu/fp7/home_en.html.
16. http://ec.europa.eu/research/fp7/index_en.cfm?pg=budget.
17. Horizon 2020 website: http://ec.europa.eu/programmes/horizon2020/en/what-horizon-2020.
18. http://ec.europa.eu/research/fp7/index_en.cfm?pg=budget.
19. http://ec.europa.eu/research/agriculture/scar/index_en.html.
20. CGIAR website: http://www.cgiar.org/ .
21. Global Alliance for Climate-Smart Agriculture: http://www.fao.org/climate-smart-agriculture/85725/en/.
22. Global Research Alliance on Agricultural Greenhouse: http://www.globalresearchalliance.org.

References

Agentschap NL (2012), *Nederlandse topsectoren in KP7 2011,* Ministry of Economic Affairs, Agriculture and Innovation. https://view.publitas.com/agentschapnl/nederlandse-topsectoren-in-kp7-2011/#page/1.

Algemene Rekenkamer (2011), *Innovaliebeleid* [Court of Auditors's report on innovation policy], September. http://www.rekenkamer.nl/Publicaties/Onderzoeksrapporten/Introducties/2011/09/Innovatiebeleid.

Alston, J. M. (2010), "The Benefits from Agricultural Research and Development, Innovation, and Productivity Growth", *OECD Food, Agriculture and Fisheries Papers*, No. 31, OECD Publishing. http://dx.doi.org/10.1787/5km91nfsnkwg-en.

AWTI (2014), *Status of the top sectors in 2014 - English summary*, Advisory council for science, technology and innovation (AWTI), October. http://www.awti.nl/upload/documents/publicaties/engels/Status-op-the-top-sectors-in-2014-Summary.pdf. [Full report in Dutch available at: http://www.awti.nl/upload/documents/publicaties/tekst/Balans-van-de-topsectoren-2014_Def_lrs.pdf].

Berkhout, P., H. Silvis and I. Terluin (2014), Landbouw-Economisch Bericht 2014, LEI-rapport 2014-013, LEI Wageningen UR, The Hague, June, 200 pp. http://edepot.wur.nl/306953.Caggiano, M. (2014), *AKIS and advisory services in The Netherlands*, Report for the AKIS inventory (WP3) of the PRO AKIS project. Online resource: www.proakis.eu/publicationsandevents/pubs.

Campi, Mercedes; Nuvolari, Alessandro (2013): Intellectual property protection in plan varieties: A new worldwide index (1961-2011), LEM Working Paper Series, No 2013/09. http://hdl.handle.net/10419/89567.

CBS Monitor (2012), *Topsectoren uitkomsten eerste meting, Centraal Bureau voor de Statistiek.* Available at: http://www.cbs.nl/NR/rdonlyres/FA8F2205-3054-41D9-A206-BC9D13F7ECBF/0/monitorentopsectorenresultatenweb.pdf.

EIM (2012), „Evaluatie EBSO 2006 2010. Effecten, doelgroepbereik en uitvoering", Zoetermeer, www.rijksoverheid.nl/bestanden/documenten-enpublicaties/rapporten/2012/04/02/hoofdrapport-evaluatie-wbso-2006-2010/hoofdrapport-evaluatie-wbso-2006-2010.pdf.

Ernst & Young (2014), *Uitkomsten feitenonderzoek matchingbehoefte op (Europese) onderzoeksubsidies*, Den Haag. http://www.rijksoverheid.nl/documenten-en-publicaties/rapporten/2014/03/13/uitkomsten-feitenonderzoek-matchingbehoefte-op-europese-onderzoeksubsidies.html.

EU SCAR (2013), *Agricultural knowledge and innovation systems towards 2020 – an orientation paper on linking innovation and research*, Brussels. doi:10.2777/3418. http://ec.europa.eu/research/bioeconomy/pdf/agricultural-knowledge-innovation-systems-towards-2020_en.pdf.

Haas, M de. (2013), *Two centuries of state involvement in the Dutch agro system*, WRR, http://www.wrr.nl/.

Innovatienetwerk (2013), *Jaaroverzicht 2013*. Available at: http://www.innovatienetwerk.org/nl/bibliotheek/rapporten/587/Jaaroverzicht2013.html.

Innovatienetwerk (2014), *Werkprogramma 2014*. Available at: http://www.innovatienetwerk.org/nl/bibliotheek/rapporten/576/Werkprogramma2014.html.

Kok de J.M.P., J. de Jonge and M. Tom (2007), *Tijdsbesteding universitaire wetenschappelijk personeel*, Den Haag. http://www.vawo.nl/wp-content/uploads/2012/06/rapporttijdsbestedingWOioOCW.pdf.

Louwaars, N, H. Dons, G. Van Overwalle, Hans Raven, A. Arundel, D.J.F. Eaton, and A. Nelis (2009), *Breeding Business. The Future of Plant Breeding in the Light of Developments in Patent Rights and Plant Breeder's Rights*, Centre for Genetic Resources (CGN), Wageningen UR.

Ministerie van Financiën (2014), *IBO Wetenschappelijk onderzoek*, May. http://www.rijksoverheid.nl/documenten-en-publicaties/rapporten/2014/05/01/ibo-wetenschappelijk-onderzoek.html.

Ministry of Economic Affairs (2013), "Summary Chart Enterprise Policies" (unpublished), ERAWATCH website,

OECD (2014a), "Analysing Policies to improve agricultural productivity growth, sustainably: Revised framework". http://www.oecd.org/tad/agricultural-policies/Analysing-policies-improve-agricultural-productivity-growth-sustainably-december-2014.pdf.

OECD (2014b), *OECD Science, Technology and Industry Outlook 2014*, OECD Publishing, Paris. DOI: http://dx.doi.org/10.1787/sti_outlook-2014-en.

OECD (2014c), *OECD Reviews of Innovation Policy: Netherlands 2014*, OECD Publishing, Paris. DOI: http://dx.doi.org/10.1787/9789264213159-en.

OECD (2014d), *OECD Territorial Reviews: Netherlands 2014*, OECD Publishing, Paris. DOI: http://dx.doi.org/10.1787/9789264209527-en.

OECD (2013), *Agricultural Innovation Systems: A Framework for Analysing the Role of the Government,* OECD Publishing, Paris. DOI: http://dx.doi.org/10.1787/9789264200593-en.

OECD (2012), *OECD Economic Surveys: Netherlands 2012*, OECD Publishing, Paris. DOI: http://dx.doi.org/10.1787/eco_surveys-nld-2012-en.

Onwezen, M. et al. (2013), *De Agrofoodmonitor. Maatschappelijke waardering van de agro & food sector,* Wageningen UR.

Park, W. G. (2008), "International Patent Protection: 1960-2005", *Research Policy*, No. 37, pp. 761-766.

Peper, B. (1996), Duurzame kennis, Duurzame landbouw; een advies aan de Minister van LNV over de kennisinfrastructuur in 2010, Den Haag.

PBL (2014), *Green gains: in search of opportunities for the Dutch economy*, Netherlands Environmental Assessment Agency. http://www.pbl.nl/en/publications/green-gains-in-search-of-opportunities-for-the-dutch-economy.

Rogers, E.M. (1962), *Diffusion of Innovations*, Free Press, Glencoe. ISBN 0-612-62843-4.

Roseboom, J. and H. Rutten (1998), The transformation of the Dutch agricultural research system; An unfinished agenda, http://www.ingentaconnect.com/content/els/0305750x/1998/00000026/00000006/art00039.

RVO (2014a), "MKB-innovatiestimulering Topsectoren (MIT)", www.rvo.nl/subsidies-regelingen/mkb-innovatiestimulering-topsectoren-mit (translated version).

RVO (2014b), WSBO/RDA programme, 15 April 2014 and correspondence with Ministry of Economic Affairs.

Schoutens, C.H.F. (2004), *De omvang van matching. Onderzoek naar de effecten van matching van 2e en 3e geldstroomfinanciering op de beleidsruimte van Nederlandse, publieke kennisinstellingen*, Ernst & Young, Den Haag. http://www.awti.nl/upload/documents/publicaties/tekst/as30.pdf.

Spiertz, J.H.J., and M.J. Kropff (2011), Adaptation of knowledge systems to changes in agriculture and society, the case of the Netherlands NJAS.

Steen, J van (2014), Total investment in research and innovation, Den Haag.

Tiemeijer W., and J.F.M. de Jonge J (2012), *Hoeveel vertrouwen hebben Nederlanders in wetenschap?* WRR, Rathenau.

Van der Meer R.W. (2014), "Innovatie in de land- en tuinbouw 2013", Wageningen, LEI Wageningen UR. http://edepot.wur.nl/317866.

Van der Meer, R.W., and M.A. Van Galen (2013), "Innovatie in de land- en tuinbouw 2012", Report LEI 13-097, LEI Wageningen UR, 20 pp. http://edepot.wur.nl/286031.

Van der Vlist, A.J., and M.A. van Galen (2005), "Innovatiemanagement en de rol van LNV; Inventarisatie van Studies en Aanbevelingen", Report LEI 6.05.10, LEI Wageningen UR, 64 pp. http://edepot.wur.nl/24267.

Van Galen, M.A., and J.A.A.M. Verstegen (2008), "Innovatie in de agrarische sector; We kunnen er niet genoeg van krijgen!", Report LEI 2008-067, LEI Wageningen UR, 83 pp. http://edepot.wur.nl/24329.

Van Galen, M.A. (2012), "Innovatie en vernieuwing in de land- en tuinbouw in 2010 gedaald", *Agri-monitor* 2012, April.

Van Galen, M., K. Logatcheva, T. Bakker, E. Oosterkamp, and G. Jukema (2013), Innovatie in de levensmiddelenindustrie; Een internationale benchmarkstudie. LEI Wageningen UR.

Verstegen, J.A.A.M., M. Klopper, and H. Schuite (2003), "Een hernieuwde kijk op individuele besluitvorming in de glastuinbouw [Individual decision making in greenhouse horticulture revisited]." Report LEI 7.03.13, LEI Wageningen UR, Oktober, 84 pp. http://edepot.wur.nl/35814.

World Economic Forum (2013), *The Global Competitiveness Report 2013-2014: Full data Edition*, Geneva 2013. http://reports.weforum.org/the-global-competitiveness-report-2013-2014/#=.

Wageningen UR (2014), *Jaarverslag 2013*. http://www.wageningenur.nl/upload_mm/f/b/7/083f3a68-912c-4413-9794-e42e65533850_Jaarverslag%202013%20ENG_3.pdf.

Wageningen UR (2013), *Annual report 2013*, Wageningen. http://www.wageningenur.nl/upload_mm/f/b/7/083f3a68-912c-4413-9794-e42e65533850_Jaarverslag%202013%20ENG_3.pdf.

List of websites consulted

http://www.nwo.nl/en

http://www.rvo.nl/subsidies-regelingen/wbso-en-rda

http://www.wageningenur.nl/en/About-Wageningen-UR.htm

http://www.vsnu.nl/

https://www.tno.nl/content.cfm?context=thema&content=innovatiegebied&laag1=891&laag2=903&item_id=903&Taal=2

http://www.nizo.com/home/

http://www.tifn.nl/

http://www.louisbolk.org/

http://www.rivm.nl/English

http://www.foodvalley.nl/English/default.aspx

http://www.lto.nl/english

http://www.frieslandcampina.com/english/innovation/innovation-centre-wageningen.aspx

http://www.dlvplant.nl/uk/content/index.html

http://www.irs.nl/en

http://www.innovatienetwerk.org/en/

https://www.vwa.nl/english

http://www.groenonderwijs.nl/feiten_cijfers_deelname_groen_hbo.php

http://nvao.com/

http://www.topuniversities.com/university-rankings/university-subject-rankings/2014/agriculture-forestry#sorting=rank+region=+country=+faculty=+stars=false+search=

http://english.rvo.nl/subsidies-programmes/wbso-research-and-development-rd-tax-credit

http://www.cbs.nl/nl-NL/menu/informatie/beleid/publicaties/papers/archief/2012/monitor-topsectoren.htm

http://www.maatschappelijkcafe.nl/events/2014/nieuwjaarsdebat

http://www.rathenau.nl/actueel/nieuwsberichten/2014/03/wetenschap-linkse-watjes-moeten-vertrouwen-verdienen-en-behouden.html

ORGANISATION FOR ECONOMIC CO-OPERATION AND DEVELOPMENT

The OECD is a unique forum where governments work together to address the economic, social and environmental challenges of globalisation. The OECD is also at the forefront of efforts to understand and to help governments respond to new developments and concerns, such as corporate governance, the information economy and the challenges of an ageing population. The Organisation provides a setting where governments can compare policy experiences, seek answers to common problems, identify good practice and work to co-ordinate domestic and international policies.

The OECD member countries are: Australia, Austria, Belgium, Canada, Chile, the Czech Republic, Denmark, Estonia, Finland, France, Germany, Greece, Hungary, Iceland, Ireland, Israel, Italy, Japan, Korea, Luxembourg, Mexico, the Netherlands, New Zealand, Norway, Poland, Portugal, the Slovak Republic, Slovenia, Spain, Sweden, Switzerland, Turkey, the United Kingdom and the United States. The European Union takes part in the work of the OECD.

OECD Publishing disseminates widely the results of the Organisation's statistics gathering and research on economic, social and environmental issues, as well as the conventions, guidelines and standards agreed by its members.

OECD PUBLISHING, 2, rue André-Pascal, 75775 PARIS CEDEX 16
(51 2015 14 1 P) ISBN 978-92-64-23845-9 – 2015

www.ingramcontent.com/pod-product-compliance
Ingram Content Group UK Ltd.
Pitfield, Milton Keynes, MK11 3LW, UK
UKHW051502290926
15471UKWH00008B/1359

9 789264 238459